TRANSFORMING THE FISHERIES

PATRICK BRESNIHAN

Transforming the Fisheries

Neoliberalism, Nature, and the Commons

University of Nebraska Press
LINCOLN AND LONDON

Manufactured in the United States of America

Library of Congress Cataloging-in-Publication Data

Names: Bresnihan, Patrick, author.
Title: Transforming the fisheries: neoliberalism, nature, and the commons / Patrick Bresnihan.
Description: Lincoln: University of Nebraska Press, [2016] | Includes bibliographical references and index.
Identifiers: LCCN 2015038336
ISBN 9780803254251 (cloth: alk. paper)
ISBN 9781496206404 (paper: alk. paper)
ISBN9780803285842 (epub)
ISBN 9780803285859 (mobi)
ISBN 9780803285866 (pdf)
Subjects: LCSH: Overfishing—Europe. |
Fishery management—Europe. |
Environmental policy. | Natural resources,
Communal. Classification: LCC SH253 .B74 2016 |
DDC 338.3/727094—dc23 LC record available at
http://lccn.loc.gov/2015038336

Set in Lyon Text by L. Auten.

For Mum and Dad

Contents

Acknowledgments

This book began from a conversation with Hilary Tovey in 2007. It was her initial support and kindness that led me to start a PhD in Trinity College Dublin, and without her ongoing patience and generosity I would never have turned that research into a book. I received similar kindness from Eamonn Slater, who supervised the writing up of the book in Maynooth University in 2014/15. Our lunches in Kilcock and conversations about Rundale and ecology were always much appreciated.

I have received funding for both my doctoral research and the writing of this book. I would like to thank Trinity College, Dublin, for the scholarship I received as an undergraduate and the AXA Foundation for funding my doctorate. To complete the book, I received an Irish Research Council Government of Ireland Postdoctoral Fellowship in 2014. I have received additional financial assistance from the Sociology Department in Maynooth. I am also grateful for the patience and care given by the editorial and design team at the University of Nebraska Press—particularly Ann Baker, Chris Steinke, and Bridget Barry.

The months I spent in Castletownbere, County Cork, doing fieldwork were some of the most interesting and rewarding I have known. It would not have been so without the generosity and warmth of the people I met there. Thanks to the fisheries managers, scientists, civil servants, and, of course, fishermen who gave their time and showed me so much. Special thanks to Mary for deciding to move down there with me and for being there through it all.

As is the case with all intellectual work, this book owes many debts to people I have come into contact with over the years. It is almost impossible to pinpoint which conversations and exchanges mattered and how, but it is their cumulative effect that has shaped and still shapes my thinking about the problems addressed in this book. So thanks to Alex Dubilet, Mark Garavan, John Holloway, Emanuele Leonardi, Jawad Moustakbal, Marcela Olivera, Christian Parenti, Ross Perlin, Maria Puig de la Bellacasa, Tom Stammers, Eddie Yuen, Alessandro Zagato, and many more. My participation in the Authority Research Network has also given me all sorts of intellectual nourishment over the past five years as well as new friendships.

Something I have always appreciated is the time and generosity given to me by teachers and individuals whose work I value. This goes all the way back to my history teacher in school, who always assumed his students were intelligent and always had time to talk. More recently, two people that stand out in a similar respect are Dimitris Papadopoulos and Peter Linebaugh. Apart from influencing my work through their writing, they have always been more than generous as people and teachers.

Closer to home, my sister, Sive, was one of the first people to read and comment on drafts of my thesis and has always given her little brother insights and support. Similarly, Naomi Millner, whose energy and work constantly inspires, has shared so many ideas with me about history, nature, and the commons. Going back to the very beginning of my PhD, my main collaborator and friend has always been Mick Byrne—he still is and I feel very lucky for that. Without a doubt, though, the person who has given the most support in the writing of this book has been Rachel O'Dwyer, who provided endless conversations, read drafts, and was generally so positive and patient. She continues to amaze through her modesty and intelligence.

Finally, I want to thank the Catholic Library on Merrion Square and the National Library on Kildare Street where I did much of my writing for the PhD and the book—I hope they never close.

TRANSFORMING THE FISHERIES

1

Introduction

Ecological Crises and Beyond

The Ghost of Malthus

In 1998 ecologist Garrett Hardin wrote a sympathetic reappraisal of Thomas Malthus's text *An Essay on the Principle of Population*, published two hundred years earlier (1998). He relates a parable that Malthus added to the second edition. In the parable, a man comes to the table of "nature's mighty feast" and asks if he can have a seat. Some of the guests have sympathy for him and make room. Immediately, other "intruders" appear demanding that they also be admitted to the feast. Malthus concludes, "The order and harmony of the feast is disturbed, the plenty that before reigned is changed into scarcity; and the happiness of the guests is destroyed by the spectacle of misery and dependence in every part of the hall, and by the clamorous importunity of those, who are justly enraged at not finding the provision which they had been taught to expect" (Malthus 1803, 531; Hardin 1998, 181). The guests thus learn the lesson that the "great mistress of the feast" already knew: they must refuse any newcomers when the table is already full.

Although this anecdote was taken out from subsequent editions, it remains a powerful metaphor for both supporters and critics of Malthusian theories of overpopulation and the "naturalization" of scarcity (Dale 2012; Mehta 2010). Thirty years before he invoked Malthus's story, Hardin had already given it new life through his own parable, "The Tragedy of the Commons." Published in 1968, Hardin's essay was only one of many stark warnings about impending social and environmental catastrophe if rapid population growth continued to put pressure on limited resources. In the same year, Paul Ehrlich's book *The Population Bomb* opened with an alarmist statement about the need to accelerate the global death rate if the problems of hunger and famine were to be averted (Ehrlich 1968). These deliberately polemical accounts found support in scientific data and predictions based on current models of development and resource availability. In 1972 the newly formed Club of Rome published their well-known report *The Limits of Growth*, and five years later a research group based in MIT released the *Global 2000 Report to the President* (Pirages and Cousins 2005). These reports showed that demands on soil, forests, fisheries, and water supplies would reach critical levels by the turn of the century. The problem was growing demand on a finite planet. Instead of just defining a crisis in production or growth rates, these reports identified a wholesale crisis in the sphere of biophysical reproduction (Cooper 2008). Pointing to the potentially catastrophic consequences of unregulated growth, writers like Hardin did not then find it hard to recall and reestablish Malthus's theories of overpopulation. The basic "law" he propounded was that the exponential growth of the human population would produce demands that would outstrip available resources. Although Malthus may have provided some unsavory solutions to this problem, the validity of his arguments remained. Failure to acknowledge this law of nature would result in far worse outcomes: "injustice is preferable to total ruin," as Hardin succinctly put it (1968, 1247).

Today, multiple and mounting ecological crises appear, if anything, to be worse than the predictions of forty years ago. Scientific evidence documents the sixth mass extinction as well as disruptions to the hydrological cycle, the soil cycle, and, perhaps most importantly of all, the carbon cycle, otherwise known as anthropogenic climate change (Kolbert 2014). The media speculates and increasingly reports on "perfect storms" of food shortages, water scarcity, and insufficient energy resources with devastating social, economic, and geopolitical consequences (Parenti 2011). Mirroring these accounts, and sometimes indistinguishable from them, are the seemingly endless stream of dystopian films and books that populate our cultural imaginary (Lilley et al. 2012). The emphasis on limits, shortages, thresholds, and overcapacity is understandable, but it also reveals a recurring Malthusian trope about the narrowing of future possibilities in the face of "natural" limits and the urgent need to reorganize society in response to these limits (Mehta 2010). As historian Iain Boal writes, "Scratch an environmentalist and you'll probably find a Malthusian" (Boal 2006).

Perhaps the clearest sign of the lingering ghost of Malthus is in the growing popularity of the term *anthropocene.* First coined in the 1980s by ecologist Eugene Stoermer, the term has spread far beyond the concerns of the earth and atmospheric sciences. It refers to a new geological period following the Holocene when humanity, "anthropos," has played a decisive and largely destructive role in geological and environmental transformations. While there is disagreement over when "we" began having such an impact on the lithosphere (see Moore 2014), the most popular periodization dates the Anthropocene from sometime around 1800, the moment coal became the principal energy source of a carbonized human civilization (Crutzen and Stoermer 2000; *Economist* 2013). The term "anthropocene" appears to both respond to and explain the unprecedented and multiple ecological crises we are currently experiencing. It is compelling because it finally appears to take seriously what we have failed to recognize for so long: we, "humanity," are responsible for the over-exploitation and degradation of "nature," and something urgent must be done to rebalance our relationship with it. With this, the current storm of ecological crises is neatly translated into a conflict between unregulated human activity and limited biophysical nature. Establishing this as the point of consensus opens a space for policy makers, scientists, companies, and citizens to work together to rebalance a system that has fallen out of sync.[1] A prime example of this transition from large-scale depletion of resources to consensus-based environmental management is the crisis of overfishing in the Irish and European fisheries, the focus of this book.

In 1844 Thomas Huxley, a leading Victorian scientist, presented a paper to the "Great International Fishery Exhibition" in London.

He claimed that "the cod fishery, the herring fishery, the pilchard fishery, the mackerel fishery, and probably all the great sea-fisheries are inexhaustible; that is to say that nothing we do seriously affects the number of fish" (quoted in Graham 1943, 111).[2] After centuries of exploitation, greatly intensified since the 1970s, the extractive demands of the fishing industry have caught up with the reproductive capacities of most commercially targeted fish stocks (Food and Agricultural Organization 2010; Rogers 1995). The project of capitalist modernity has finally conquered the deep-sea fisheries once thought to be inexhaustible (Campling et al. 2012; Clausen and Clark 2005). This conquest encapsulates a familiar history of capitalist development that was far from "natural" or linear. The modernization of fishing fleets, the development of onshore landing, processing, and distribution infrastructure, and the opening up of new global markets required political and economic investments that excluded other modes of marine production, knowledge, and culture.[3] Now, in place of Thomas Huxley's nineteenth-century optimism, we are more likely to encounter the catastrophist claims of someone like Charles Glover, an environmental campaigner whose book *The End of the Line* was turned into a popular documentary in 2007. Released in cinemas as part of a wider media campaign to inform the public about overfishing, the film, as the title suggests, is part of a new genre of eco-catastrophe. Combining footage of industrial-scale fishing with clips of international scientists predicting the future collapse of global fish stocks, its message is unambiguously stark: if the level of fishing does not reduce dramatically, the oceans will be emptied.

The crisis of overfishing is particularly severe in the European fisheries. In 2008 the International Council for the Exploration of the Sea (ICES) concluded that 35 of 41 commercial fish stocks in European waters were overfished, compared to 25 percent of fish stocks worldwide (Commission for Environmental Cooperation 2008b). Another report estimated that the European fishing industry exceeded sustainable fishing levels by 40 percent (Commission for Environmental Cooperation 2008a). As I was told countless times during my research, the problem stemmed from the fact that "too many fishermen were chasing too few fish." The European Commission has described this as a "vicious cycle" as fishermen are pressured into fishing more intensively in ever more distant fishing grounds to repay debts and compete with fishermen in other parts of the world.

As with other environmental issues, the crisis of overfishing is a growing concern not only for policymakers but also for environmental campaigners, nongovernmental organizations, and the public.[4] Growing awareness among scientific, fishing, and environmental communities about declining fish stocks and the unsustainability of the fishing industry have now pushed questions of conservation and sustainability into the center of debates over the future management of the Irish and European fisheries.

At "Back to the Future," a conference held in Dublin with environmental NGOs in June 2011, the Irish Minister for Agriculture, Food and the Marine, Simon Coveney, said that fisheries management could not return to the past, to a time when many people made a living from the sea, when there were healthy fish stocks and abundant biodiversity. There was "no choice but to become a modern fishery adapted to global realities" (Coveney 2011). Organized as part of a lobbying campaign for the upcoming reform of the European Common Fisheries Policy (CFP), the premise of the meeting was to imagine a way beyond the current crisis, "restoring" the fisheries and fishing communities to a sustainable path. Minister Coveney made it clear that this would require two strategies: first, taking "courageous decisions" to limit fishing effort; and second, developing new economic opportunities around "green" growth and sustainability. Minister Coveney pinpoints what is required to move beyond the crisis of overfishing: a transformation in the mode of production, a shift away from the unsustainable extraction of limited marine resources to a mode of production based on "green" values. European fishing policy reflects this ambition with member states setting themselves the target of reducing fishing mortality by 50 percent in some fisheries by 2020. The scale of this task is vast. As the European Commission makes clear, achieving this will require "*radical changes to the way Europe's fisheries are managed*—changes which will reverse economic and institutional incentives to overfishing and replace them with a system which positively encourages good stewardship of our oceans and seas by all those who live from them" (Commission for Environmental Cooperation 2008a, 7; emphasis added).

To those critical of attempts to "green" capitalism, this aspiration toward a sustainable fishing industry might appear naïve at best and meaningless rhetoric at worst. In familiar Malthusian terms the phenomenon of overfishing is "naturalized," becoming the fact around which fisheries managers, policy makers, and scientists can come together to negotiate and work out pragmatic and measurable solutions. The only questions posed in this account are how to ensure the continued biological reproduction of fish stocks ("Nature") and how to provide new opportunities for the creation of profits ("Capitalism") (Kenis and Lievens 2014).[5] This obscures one of the critical insights of Marxist analysis and, more recently, political ecology (Blaikie 1985; Moore 2003; Peet and Watts 1996): scarcity or the degradation of ecosystems is not "natural" but the result of specific, uneven, and contestable processes of *social* production. In neglecting this we are prevented from asking "the politically sensitive, but vital, question as to what kind of socio-environmental arrangements do we wish to produce, how can this be achieved, and what sort of natures do we wish to inhabit" (Swyngedouw 2007, 23).

But what does this critique tell us about how these dominant, bioeconomic narratives are reshaping the interactions between society and nature? What do Coveney's seemingly benign words mean when translated into new scientific, economic, and regulatory practices in fisheries management today? Is it just "business as usual," or are these cumulative efforts to manage ecological crises such as overfishing giving rise to *new* ways of knowing, valuing, and producing nature? These are important questions for understanding what is at stake in contemporary environmental governance: as "nature" transforms from being a raw material for extraction to something that must be cared for and valued within a "green" economy, certain ways of knowing and doing will be counted as "productive" and "environmental" and others will not. At its heart then, this process is about the redrawing of boundaries, the generation of novel forms of inclusion and exclusion. This book examines these transformations and the different ways they are being justified and implemented.

In this sense the book also tries to take Malthus seriously. "Malthusianism" has meant different things at different times, but a common understanding is that Malthus was politically and morally conservative, advocated all manner of forced population control, and favored natural checks on the poor such as famine and disease (see Mayhew 2014; Ross 1998). He is commonly cast, and thus dismissed, as an ideological advocate of the elites, a high priest of capitalist enclosure (Dale 2012). There is a general (and justifiable) tendency to focus on the negative social consequences of his thinking and the policies he inspires: "Somebody, somewhere, is redundant, and there is not enough to go round," as David Harvey rightly concludes (Harvey 1974, 273). However, while his ideas and writings undoubtedly help to justify social inequality, defining him solely as an apologist for particular interests misses the real force of his analysis and thus limits our capacity to effectively move beyond it. A different reading situates Malthus more broadly within the liberal current of thought that emerged during the eighteenth century in Britain and elsewhere (Mayhew 2014; Winch 2013). This historical framing also reminds us that Malthus and others were responding to particular social and material conditions: at the end of the eighteenth century, there were real and urgent problems of food scarcity and associated social and political upheavals. Malthus was part of a generation of thinkers that began to problematize such crises in a radically different way. I trace how the force of this liberal reasoning still operates today through the management of ecological crises such as overfishing. In the following section, I will briefly outline how my book contributes theoretically to the study of ecological crisis and transformation and where it sits within existing debates on neoliberalism and nature.

Neoliberalism and the New Enclosures

Over the last thirty years different biophysical resources in more and more parts of the world have been subjected to processes of commodification and privatization (Castree 2008a, 2008b). Continued and expanding commodity production has fed demand for raw materials, including land, water, and energy. This expansion has given rise to "classical" forms of enclosure, such as widespread land grabs in the Global South (Heynen et al. 2007; McMichael 2011). However, new forms of environmental management have also turned to the market to achieve its goals: mounting environmental problems at regional and global scales, the inability of existing state institutions to deal with them, and new commercial opportunities arising from the "green" economy have all promoted the embrace of market-based instruments for managing environmental problems such as overexploited resources, pollution, or habitat destruction (Heynen and Robbins 2005; Mansfield 2004; McCarthy and Prudham 2004). In water resource management, not only has the private sector become more involved in water services but the extension of the user-pays principle reflects the normalization of economic values when it comes to resource allocation (Bakker 2003, 2005; Budds 2004; Kaika 2003; Smith 2004); managers of global fisheries have introduced individual transferable quotas (ITQs) that effectively facilitate new markets in fish quotas (Mansfield 2007a, 2007b; St. Martin 2000, 2007); there are ongoing efforts to address the problems of climate change through carbon markets (Bond 2012; Lohmann 2009; Leonardi 2012); and Natural Capital accounting and Payments for Ecosystem Services (PES) provide a seemingly limitless field for commercial opportunities in the areas of biodiversity conservation (Büscher et al. 2012; Sullivan 2013). This general and multifaceted process of marketization in the area of environmental governance has led some scholars to describe it as the "neoliberalization of nature" (Heynen et al. 2007).

Unsurprisingly, the dominant critical response to the resurfacing of the market as a response to ecological crises has come from the field of Marxist political economy. In many ways, it echoes Marx's original critique of liberal economics and the faith in laissez-faire policy-making. For Marx, the "free market" was anything but free when it forced people to sell their labor and relied on the state to introduce and uphold the rule of private property. Marx described the violent process of expropriation that separated the mass of the population from necessary and direct access to the means of social reproduction (land, rivers, forests) as "primitive accumulation" (Marx [1867] 1990). This separation took place during the seventeenth and eighteenth century in Europe, forcing the population from the country into the cities, where they had to sell their labor in return for wages. For Marx, liberal economists only provided a one-sided, ahistorical account of this process, elevating the capitalist mode of production from a historically specific organization of labor and nature into a universal one. This provided the ideological defense of capitalist interests and the justification for new rounds of accumulation in response to periodic crises: the problem of scarce resources could always be displaced and overcome through the enclosure and commodification of new frontiers (Moore 2014a, 2014b).[6]

Since the telling intervention of the Midnight Notes Collective, scholars and activists have recognized that this dynamic relationship between crisis and capitalist expansion is at the heart of contemporary forms of neoliberal globalization (De Angelis 2007; Jeffrey et al. 2012; Midnight Notes 1990). This updates orthodox Marxist analysis, which understands primitive accumulation as a *particular* phase in the historical emergence of capitalism. The result is a growing literature on new forms of primitive accumulation or "accumulation by dispossession" that have arisen in response to the crisis of environmental limits (Bond 2012; De Angelis 2001; Harvey 1996).

Although this historically informed analysis of the relationship between capitalist crises and socio-ecological crisis is critically important, it often assumes that neoliberal policies—or policies that open up potentially new markets—are the result of elite, capitalist interests, rather than the result of a particular form of economic reasoning that is intimately tied up with, but irreducible to, capitalist expansion. This dismissal of neoliberal reasoning as an important site of inquiry in itself is indicated by a tendency to focus on the negative social and environmental effects of neoliberal policies rather than the logic and practice that enable neoliberalism to have such a hold over our lives (Dardot and Laval 2013; Lemke 2002, 2011a). Political economy approaches thus appear to privilege the expansionist logic of capitalism at the expense of examining the regulatory, institutional, and discursive processes that enable this growth to happen. The effect of this analysis can appear contradictory: it lends "neoliberalism" considerable coherence and power but also takes it largely for granted. Although neoliberal policies *are* creatively responding to crises of capitalist production, they are doing so in ways that are constitutive of new "natures" that go beyond a narrow concern for economic productivity, labor, and profits. In other words, it is important to take seriously what neoliberalism in the realm of knowledge-production and subject-formation is doing rather than assuming that policy makers, scientists, and even environmentalists are naively presenting (again) a one-sided account of ecological problems that effectively naturalize and universalize capitalist relations of production.

Partly in response to this idealization of neoliberalism, a body of work has emerged over the past ten years that seeks to empirically analyze processes of neoliberalization within particular geographic and institutional contexts (Castree 2008a, 2008b; Chazkel and Serlin 2010; Heynen and Robbins 2005). Although still situated within the theoretical tradition of Marxist political economy (and political ecology), this work tends to emphasize the diverse implementation of neoliberal rationalities in practice. Importantly, these scholars make clear that neoliberalism is not uniform or pure but adapts to the many different institutional settings and sociomaterial realities where it is deployed (Bakker 2010; Fine 2009). These efforts to describe and examine "actually existing neoliberalism" (Brenner and Theodore 2002) emerge in response to the gap between the ideological claims and representations of neoliberalism and the complex, messy, even contradictory ways it materializes in the world. The emphasis of this work is thus on examining the *process* and *practice* of neoliberalism; rather than defining neoliberalism as a coherent set of institutions or a program to shape reality, these scholars argue that neoliberalism is better understood as a hybrid process from the beginning (Peck 2004). This approach reveals the heterogeneity of neoliberalism, the different institutional contexts in which it takes place, and the involvement of both state and nonstate actors in shaping its development (Larner 2000, 2003; Mansfield 2004, 2006, 2007b). As Noel Castree has argued, however, empirically rich case studies that challenge the supposed purity of neoliberalism can also undermine our capacity to grasp the force of neoliberalism as the governing rationality of contemporary life, particularly as it relates to managing ecological crises.[7]

Largely missing in accounts of the "neoliberalization of nature," Michel Foucault's historically rooted analysis of liberal and neoliberal thought can help respond to this apparent impasse between ideologically "strong" and empirically "weak" conceptualizations of neoliberalism.[8] His understanding of liberalism cannot, however, be separated from the analysis of biopower developed in his later work. In the last of his 1976 lectures at the College de France and in his book *The History of Sexuality,* Foucault outlined how until the end of the eighteenth century, sovereign power was characterized by a power of "deduction": the legal deprivation of goods, products, services, and, in extreme cases, life itself from political subjects. In contrast, biopower is characterized by a power of "production" that seeks to administer, develop, and foster life, "a power bent on generating forces, making them grow, and ordering them, rather than one dedicated to impeding them, making them submit, or destroying them" (Foucault 1998, 136).[9] With this shift, nature is no longer understood to be "external, holy and unchangeable" but rather consists of "natural processes of life" that are subject to measurement and regulation (Lemke 2010). The task of government was to better understand these underlying processes in order to shape and channel them toward certain "common" goals, such as increased economic output. This new relationship between knowledge and power gave rise to a new liberal "art of government" or governmentality, as Foucault calls it, that is not concerned with "imposing law on men but of disposing of things: that is of employing tactics rather than laws, or even of using laws themselves as tactics—to arrange things in such a way that, through a certain number of means, such-and-such ends may be achieved" (Foucault 1991, 95). These "tactics" do not derive from a preexisting authority but are an ongoing response to particular social and environmental phenomena. It is precisely this pragmatism, this situated-ness, this refusal to engage with "political" or "ethical" questions that makes liberal forms of government so effective at managing the population. As Thomas Lemke writes, the "perspective of governmentality makes possible the development of a dynamic form of analysis that does not limit itself to stating the 'retreat of politics' or the 'domination of the market' but deciphers *the so-called 'end of politics' itself as a political programme*" (Lemke 2000, 10; emphasis added).

It is no coincidence that Foucault traces the emergence of biopower, a power over life itself, to the second half of the eighteenth century and the culmination of so-called primitive accumulation, which Marx identified as being so central to the historical transition toward capitalism. Although Foucault does not refer explicitly to the enclosures and "improvements" unfolding across both Britain and France at this time, his analysis can and should be read as an important complement to Marxist analysis. The second half of the eighteenth century witnessed "improvements" in land-use, husbandry, and agricultural production that went hand in hand with the enclosure of the open-field system and the diverse commons of land, forest, and river (Barrell 2010; Linebaugh 2008, 2011; Mayhew 2014; Neeson 1996; Thompson 1993).[10] Important scholarly work by Carolyn Merchant, Silvia Federici, and Jason Moore has shown the intimate connections between the historic emergence of capitalism and the disciplining and control of the sphere of "reproduction"—those situated forms of knowledge, practice, and value that were necessary for the direct and ongoing reproduction of collectives of human and nonhuman life (Federici 2001, 2004, 2012; Merchant 1980; Moore 2014a, 2014b). These accounts reveal how the history of enclosures and "improvements" is not just a history of material dispossession. It is also a history of the exclusion of certain ways of knowing and relating to the land, forests, rivers, and animals.[11] Foucault's work allows us to connect the violent appropriation that characterized primitive accumulation with the emergence of a more general, productive regime of knowledge-power (Goldstein 2013). It is important to make this connection today as new modes of capitalist accumulation develop alongside new ways of knowing and organizing ecological and life processes (Barca 2007; Federici 2004; Merchant 1980; Moore 2003; Nealon 2008; Sullivan 2013). As Foucault writes,

> Biopower was without question an indispensable element in the development of capitalism; the latter would not have been possible without the controlled insertion of bodies into the machinery of production and the adjustment of the phenomena of population to economic processes. . . . . The adjustment of the accumulation of men to that of capital, *the joining of the growth of human groups to the expansion of productive forces and the differential allocation of profit, were made possible in part by the exercise of biopower in its many forms and modes of application.* (Foucault 1998, 140–41; emphasis added)

Although governmentality studies tend to focus on the micro-practices of the state, it is important to recognize that underlying ideas and assumptions about human and nonhuman nature lie at the heart of liberal modes of government. What distinguishes these forms of world-making, however, is that normative ideas are not exactly imposed on reality as prescriptive norms but are more accurately verified through the activity of governing itself, through the gradual composition of a reality that is rendered amenable to calculation and regulation. Throughout my interviews with fisheries scientists, fisheries managers, and policy makers, I encountered a consistent commitment to making policies work "on the ground." Their analysis and recommendations emerged through prolonged engagement with the socioeconomic and ecological dimensions of the fisheries they were working with. This iterative and highly reflexive process reflected a commitment to identifying and including *more* aspects of reality in order to achieve the "common" (and measurable) goals of economic and environmental sustainability. The success or failure of a particular policy was judged not in terms of its "fairness" (or according to some other normative criteria) but solely according to its measurable effects vis-à-vis these already defined goals. There was never any overarching, authoritative plan for the fisheries, and it is this openness that makes neoliberal nature-making so difficult to contest. In this sense, neoliberalization is better understood as an activity (rather than a set of institutions) that both responds to and shapes the different social and ecological contexts it operates in.[12]

Although the making of neoliberal natures may not be "true," this does not mean it is not real. As social theorists Pierre Dardot and Christian Laval write, "neo-liberalism is not merely destructive of rules, institutions and rights. It is also *productive* of certain kinds of social relations, certain ways of living, certain subjectivities. In other words, at stake in neo-liberalism is nothing more, or less, that the *form of our existence*—the way in which we are led to conduct ourselves, to relate to others and to ourselves" (Dardot and Laval 2013, 8). Over two hundred years ago, Arthur Young, an eighteenth-century agronomist and "improver," had a similar insight: "if you go into Banbury-market next Thursday you may distinguish the farmers from enclosures from those from open-fields; *quite a different sort of men*; the farmers are as much changed as their husbandry—quite new men, in point of knowledge and ideas" (Young cited in Barrell 2010, 71; emphasis added).[13] This simple observation captures how liberal forms of government transform individuals from targets of government policies into *instruments* of these policies, the means of achieving certain environmental and economic goals. This is evident in the context of transformations in fisheries management and environmental governance more generally.

In the past, fishermen were literally outside the reach of power, operating in the open seas, able to avoid what regulations existed through their peripheral location and the absence of any effective policing. This is commonly expressed in policy reports as noncompliance, a phenomenon borne out of a lack of regulation, control, security, and representation. The European Union has thus identified a new "culture of compliance" as the only way of achieving a sustainable fishing industry:

> Without active collaboration between them [industry and managers], even the best-drafted regulations founded on the best-researched science, and supported by carefully targeted subsidies can achieve little. Policy is only as good as its implementation. And in the final analysis, it is the people who work in the fishery who have to make that policy a reality, *by adopting it fully in their daily practice*. (Commission for Environmental Cooperation 2008a, 9; emphasis added)

Fishermen are no longer the passive and largely absent subjects of distant management and regulation but "active" participants in the management of the fisheries (Mikalsen and Jentoft 2001; Kooiman et al. 2005; Österblom et al. 2011; Wiber et al. 2004). Regulators see the inclusion of users in resource management not only as a response to weaknesses in previous institutional models but also as a more effective and productive form of governance (see Acheson 2003; Baland and Platteau 1996; Olsson et al. 2006). There is a growing acceptance that fisheries *cannot* be managed without the active support of fishermen (Jentoft and McCay 1995, 1996; Kooiman et al. 2008; Pomeroy 1997). The path toward the "sustainable" fisheries of the future does not therefore simply involve the exclusion of fishermen. Instead, it demands new forms of inclusion that reshape how individuals act and how they relate to other people and to their environments. These new forms of environmental governance work by incentivizing and inciting fishermen to relate to others and the worlds they inhabit in particular ways. As scholar Arun Agrawal writes, "policies aiming at greater decentralisation and participation are about new technologies of government. To be successful, they must redefine political relations, reconfigure institutional arrangements, and transform environmental subjectivities" (2006, 7). In this book I hope to show how ongoing efforts to manage and resolve the crisis of overfishing involves such far-reaching transformations, the production of new environmental subjects capable of operating within new and distinct ecologies and economies.

Overview of the Book

The geographic, social, ecological, and economic diversity of the Irish and European fisheries helps us trace the iterative relationship between particular contexts and problems and the forms of government that emerge around and through them. The current management of the fisheries shows how European, national, regional, and local levels of government interact with nonstate actors (NGOs, multinational corporations, local business), consumers, and producers in different ways, challenging and reconfiguring the boundaries of public, private, and civil society (Swyngedouw 2004). The first three chapters illustrate these interactions by examining three different forms of environmental governmentality.

In chapter 1, I begin with an overview of how critical responses to overfishing over the past decade have culminated in the recent reform of the European Common Fisheries Policy (CFP) in late 2012. The new CFP has stated the European Union's commitment to "returning" to a stable cycle of production in harmony with the "natural" cycle of biological reproduction. This is expressed in the goal of achieving Maximum Sustainable Yield (MSY) in all fisheries by 2020. MSY is the highest number of fish that can be taken safely each year while maintaining maximum productivity of fish populations. The question facing European and Irish policy makers is how to achieve this goal when its management system, according to a report commissioned by the European Union (EU), has presided "over an unparalleled period of decline for Europe's fishing industries" (Symes 2007, 49). This report criticizes a top-down, command-and-control, bureaucratic system that does not reflect the reality of the fisheries nor respond to the needs of fishermen.

The anachronistic, inflexible nature of the previous management system is exemplified by the policy of fixed quotas. Established in 1983 as a way of equitably dividing up fish stocks between member states and fishermen, the fixed quota system guaranteed each fisherman a share of the marine resources. However, the expansion of fishing capacity and the difficulty of matching fish catches with predetermined quotas has encouraged the controversial practice of dumping or discarding large quantities of non-quota fish at sea. In response, the new CFP has banned the practice of discarding. Although the ban appears to make economic and environmental sense and has met little resistance, it has far-reaching consequences for the fisheries and the way marine resources will be allocated. Arguably, the most controversial outcome is that the ban will require the introduction of individual transferable quotas (ITQs). The ITQ has been a controversial policy option because it effectively transforms access rights to the fisheries into assets that can be rented, bought, or sold.

The rationale behind the ban on discarding and the introduction of ITQs resonates with arguments that emerged in France and Britain in the late eighteenth century concerning the problem of grain scarcity. In a lecture from 1977, Foucault shows how these arguments were a key point in the development of a new relationship between knowledge and power, a relationship that gave rise to a new *liberal* mode of government. This development centered on the idea that the problem of grain scarcity was *real*, and that any effective analysis and policy making must begin from this assumption and not from a denial of this reality. In this chapter, I develop a connection between this early liberal perspective on the question of scarcity and the contemporary response to discarding and overfishing more generally in the European fisheries.

Although the ban on discards and the introduction of ITQ are the most significant policy changes in the CFP reform of 2012, the Irish government, European Union, and other transnational actors are designing other strategies that move beyond a classically liberal analysis of overfishing. Chapter 2 examines how the cultivation of new "green" opportunities in the fisheries sector shifts the focus of management from limited, marine resources to the environmental performance of fishermen and the market value that can be leveraged from it. In this chapter, I examine a particular project initiated by Bord Iascaigh Mhara (BIM), the Irish state agency responsible for developing the fishing industry, called the Environmental Management System (EMS). Based on similar programs in Australia that manage environmental impacts in agriculture, forestry, and fisheries, the EMS is essentially an auditing technology that allows fishermen to monitor their activities. It lets them "prove" their environmental credentials so that they can engage with regulatory authorities and gain accreditation for their catch. Eco-label accreditation aims to provide fishermen with greater access to markets and potentially higher prices for their catch. In this logic there is no contradiction between economic profitability and environmental sustainability.[14] As I show, however, the challenge of measuring and demonstrating environmental performance involves not only new auditing technologies but also highly uneven networks of transnational governance. Instead of competing with one another over limited marine resources, fishermen vie to measure, manipulate, and represent their environmental performance within these networks. I draw on Foucault's illuminating account of the distinction between liberal and neoliberal rationalities of government to establish what is at stake in this new logic of the "green" economy, specifically how it differs from liberal (and Malthusian) preoccupations with "natural" limits.[15]

The market-based tools described in chapters 1 and 2 both fit within familiar analyses of neoliberal policy-making. Chapter 3, however, examines another mode of governance that is often understood as an alternative to such policies. Community-based resource management is described as a "third-way" approach that moves beyond the limited choice between state-centered and market-based options. It is promoted particularly as a more effective form of environmental governance in small-scale, "traditional" spheres of resource exploitation (Berkes and Pomeroy 1997). Since the 1990s in Ireland, representatives of inshore fishermen and certain fisheries scientists have raised the threat of overexploitation in the inshore fisheries, particularly of lobster, the main commercial species. Over the past decade, through BIM, the Irish government has responded by attempting to introduce a framework for community-based lobster management.

In chapter 3, I examine the origins of community-based management and how rather than providing an alternative to liberal reasoning it represents a *different form* of this reasoning. It assumes the same liberal starting point as more dominant approaches to fisheries management: overfishing is the result of self-interested individuals over-exploiting a limited marine resource. What distinguishes the community-based approach is that it seeks to avoid the "tragedy of the commons" by devising institutional frameworks that are more flexible and adaptive than either private property regimes or direct state control.

My critical orientation toward the construction of new environmental subjects—as described in the first three chapters of the book—meant that the initial focus of my research was not on the experiences of fishermen themselves but rather on how new forms of fisheries management were attempting to mold the activities of fishermen within the new policy contexts of sustainability and the "green" economy (Amit 2003; Gupta and Ferguson 1997; Law 2004; Marcus 1995).[16] Instead of grand diagnoses, these new assemblages of knowledge-power required an ethnographic mapping that concentrated on "the little mutations" taking place on the ground (Ong and Collier 2005). I first envisaged my relationship with Castletownbere, the commercial fishing port in the South West of Ireland where I was based for sixteen months, as a site for the exploration of broader, cross-cutting (that is, global) tendencies in environmental governance and capitalist production.[17]

In the first three months of my research, I interviewed civil servants working in the Department of the Marine based in Clonakilty; National Parks and Wildlife Service staff in Glengarriff; researchers from the Beaufort project in Haulbowline, Cork; marine scientists in the Marine Institute in Galway; social scientists from the University of Galway; BIM development officers working on the ground in both Castletownbere and Dublin; representatives from the Castletownbere Fishermen's Cooperative; the owners of a seafood processing factory; and representatives from the Sea Fisheries Protection Authority and the South and West Fisheries Producers Organisation based in Castletownbere.

I carried out long interviews with these people in their offices or places of work. Interviews arranged through institutional connections were relatively formal. The BIM headquarters, for example, clearly set aside time for several employees to meet, present their work, and discuss particular issues raised in my email of inquiry. Other interviews, mainly those that took place in Castletownbere, arose out of informal contacts with people involved in the management of the fisheries (scientists, fisheries officers, civil servants) who passed on contact information of colleagues in other offices or agencies.

However, I soon came to realize during this fieldwork that often the only ground I was covering was that which was opened up for me by the institutional actors I was trying to critique.[18] By tracing the visible, articulate subjects and discourses that make up the world of fisheries management, I was replicating them in a way that did not seem particularly critical.[19] While it was important to document and describe these new modes of environmental governance and the ways they were modifying, including, and excluding fishermen, my research was not able to see anything outside them. Although I wanted to complicate the idea of "fixing" a research subject, I found myself doing just that, "fixing" the power relations and knowledge practices that were generating new economic and environmental subjects (see Li 2005, 2006).

Critical social theory tends to reinstate power relations by positioning the subject as nothing more than an epiphenomenon, "the outcome of a complex constellation of textual, material, institutional, historical factors" (Blackman et al. 2008). This understanding can end up dissolving subjectivity into a variety of generative models (culture, *habitus*, assemblage, or apparatus), ensuring that experience operates only in "ensnared spaces"[20]:

> Governmentality theory is contradictory: it suggests that experience is discursively constituted, but it critiques the attempt to research experience on the basis that it can only invoke experience as fixed, a given. The cost of jettisoning a close examination of the particulars of subjectification (researching lived experience is one way of doing this) is to deter engagement with the problem of alternative modes of political engagement. (Stephenson 2003, 141)

Theorists Dimitris Papadopoulos and Niamh Stephenson have sought to address the loss of experience in some contemporary critical research by turning to the everyday, continuous experience that unfolds outside modes of representation, which is "*imperceptible*" (Papadopoulos and Stephenson 2006). Instead of focusing on how certain aspects of our experience are represented and made to work within contemporary forms of governance, Papadopoulos and Stephenson thus attend to the ways everyday experiences escape such representation. A methodology of imperceptible experience must turn away from an exclusive analysis of "optic" strategies of inclusion and articulation (such as my interviews with "experts" and analysis of new tools of participative management) to the everyday "haptic" strategies employed by people in the often mundane navigation of work and life. The haptic describes the unspectacular experiences of people as they unfold around places, people, animals, and artifacts. They are thus unspectacular in two ways: neither extraordinary nor necessarily sensible within the "optic" regimes of inclusive governance.

This reconceptualization of subjectivity as something situated, relational, and distributed across time and space helped me address a challenge I faced during my research: the recognition that my "critical" concern for new forms of environmental governance was eclipsing the richness of the immediate and everyday social and material worlds that existed in Castletownbere (Gibson-Graham 2006; St. Martin 2007, 2009). In this approach there was no sense of how people in Castletownbere might act at a distance from dominant economic and governmental rationalities, not through explicit resistance but through practical forms of world-making that relied on and constituted different ways of knowing and doing.

My experiences living and working in Castletownbere and knowing its people, houses, roads, boats, land, and sea are hard to describe.[21] In the words of Julie Katherine Gibson-Graham, it was a place "which was not fully yoked into a system of meaning, not entirely subsumed to and defined within a (global) order" (Gibson-Graham 2005, xxxiii). Although I left Castletownbere to conduct interviews and return to Dublin every so often, I soon became bound up in these everyday social relations and networks. Simply by being there, I developed attachments and relationships with particular places and with people who I would meet during the course of daily interaction on the road, in the pub, or in the shops. It was through such chance relationships that I finally managed to get out to sea, mostly on the small, inshore boats that go out for a day at a time catching crab, prawns, and lobsters. Going out to sea for the first time and the "thickening" of my relationship with this place and the people who lived there opened up a different orientation in my research. I came to see fishermen from a different perspective: although they were embedded within capitalist relations of production to different degrees, exploiting fish stocks to make a living, they were also part of different social relations and practices that unfolded through the places, people, and things that populated everyday life. Here was a different value to ethnography: the tracing not of global assemblages of power and capital but of rich, situated worlds that are more palpable than they are representable (Connelly and Clandinin 2000).[22] Although these social relations and practices may be minor, and largely invisible to an outside observer, they provide a different starting point for thinking about the organization of human and nonhuman life in a context of mounting ecological crises.

In the final chapter, I draw together ethnographic accounts from my time in Castletownbere with theoretical insights from anthropology, feminism, and posthumanist social theory to discuss what I call the more-than-human commons. At the heart of this concept is the activity of *commoning*, an ongoing exchange between humans and nonhumans that is grounded in the immediate and intimate understanding that the world is *shared*. This does not spring from any idealized conception of how things should be, but from the everyday social and material needs that exist in a place like Castletownbere: it is precisely because an inshore fisherman operates in such an unpredictable, precarious environment, beyond the clock and the set wage, that he must rely on people and access to a diversity of resources to sustain himself (Van der Ploeg 2008; Scott 1998). At the same time, the interdependence between people and the place they live and work is not fixed or automatic. There is not a set stock of "natural" resources on one side and a community of human, economic agents on the other, but a dynamic, widely distributed mesh of social and natural, material and immaterial resources that are co-produced and circulate among those who participate in the making of the commons.

Understood from this perspective, the more-than-human commons, and the activity of commoning that produces and sustains it, allows us to break with the tragic, liberal distinction between passive, biophysical resources and the economically rationalizing human subject. In contrast to *homo economicus* or the neoliberal entrepreneur carving a linear path through the world, the subject of the commons is always part of "thick" interdependent relations with human and nonhuman others. This multiplicity and diversity is integral to the vitality of the commons; commoning involves the ongoing interweaving of human and nonhuman capacities to create a world that is lively and abundant, not passive and scarce. Drawing together the concepts of care and reciprocity, however, I show that the more-than-human commons still operates within limits, limits that are not fixed but worked out through negotiation and experiment. The intention of the final chapter is not to romanticize a fishing community or deny the existence of capitalist relations of production but rather to identify these "invisible" forms of commoning and the ways they constitute different subjectivities, value, and knowledge. This analysis carries on the important work of feminist scholars who politicized the sphere of reproduction not as a "natural" phenomenon but as something that is always social and collective; breaks the hold of liberal, humanistic epistemologies that individualize human subjectivity and exclude the nonhuman from world-making processes; and finally, reveals the always changing and contested relationship between the commons and enclosure.

2

The End of the Line

Scarcity, Liberalism, and Enclosure

Reaching Consensus

After all night talks on May 30, 2013, the fisheries commissioner for the European Union, Maria Damanaki, held a press conference in Brussels to announce the finalization of the new European Common Fisheries Policy (CFP). She stated, "This is a historic step for all those involved in the fisheries sector. We are going to change radically the way we fish in the future" (quoted in Commission for Environmental Cooperation 2013). Simon Coveney, acting fisheries minister under the Irish presidency of the EU, echoed her words: "Whether you're an environmentalist or a member of an NGO who has been campaigning on this issue you can be very happy with the result. It's real, it's measurable, it's based on science . . . *it's very real*" (Coveney 2011; emphasis added).

Nor was this note of optimism limited to high-ranking politicians: Hugh Fearnley-Whittingstall, the leader of Fish Fight, one of the best publicized and popular campaigns lobbying for change in European fisheries policy, hailed the deal as a "tremendous achievement" (BBC News 2013). Even Greenpeace managed a grudging congratulation: "For all its loopholes and sluggish timelines the policy has the potential to turn Europe's destructive and oversized fishing industry into a sustainable, low-impact sector" (BBC News 2013). Media organizations reporting the story were unanimous in their support. One journalist in a liberal UK newspaper went so far as to describe it as "a great example of an increasingly democratic EU beginning to work rather well" (Hutton 2013).

One of the reasons the new CFP received such widespread praise was its commitment to putting the sustainability of fish stocks at the center of policy and decision-making.[1] This can be traced back to the European Commission's consultation process for the new CFP in 2008. For the first time in the thirty-year history of the CFP, the emphasis was on returning to a "natural" balance between fishing effort and available fish stocks:

> Most often, the only way to get back into a virtuous circle is to give nature the time and space to do her work. In other words, while economic and social sustainability are core goals of the Common Fisheries Policy (CFP), ecological sustainability is necessarily more fundamental: for it is the biological cycle of reproduction and renewal which determines whether the human activities based on it are sustainable or not. (Commission for Environmental Cooperation 2009, 8)

The recent reform of the CFP (2013) has enshrined this principle in the overarching goal of achieving Maximum Sustainable Yield (MSY) in all commercial fisheries by 2020. MSY is defined as the highest catch that can be safely taken annually while maintaining the size and maximum productivity of the fish population. This number is set each year by the International Council for the Exploration of the Sea (ICES) in concert with marine scientists from relevant national governments. The MSY is both a target and a starting point for fisheries management across Europe; it is a "common" goal around which scientists, fisheries managers, consumers, and fishermen can work together to achieve *real, measurable* changes in how the fisheries are exploited and managed. This apparent consensus has also prompted far-reaching and critical reflection on the structural failings of existing fisheries management and the need for radical changes in how the problem of overfishing is understood and what new approaches are required.

The consultation document from 2009 states bluntly that the existing model of European fisheries management was *incapable* of bringing about the necessary "wholesale transformations" required to conserve fish stocks (Commission for Environmental Cooperation 2009)—what Joe Borg, the fisheries commissioner at the time, described as a shift from the "vicious cycle" of the past to the "virtuous cycle" of the future. Recent experience has shown that stronger, direct regulation does not have the desired effects on the fishing industry and involves considerable costs to the member states and European Commission. The shift to a "virtuous cycle" will thus require greater collaboration between the fishing industry, fisheries managers, and scientists. This is deemed necessary to improve the quality and quantity of data about the fisheries and to devise fisheries-specific programs that are workable and produce measurable results in terms of reducing the levels of fish extraction. Commissioner Damanaki has articulated this explicitly in terms of devolving managerial tasks "down" to the fisheries: "So, we need to enhance the aspect of de-centralized decision-making. *Issues that are merely technical do not belong to the political level*" (Damanaki 2011d, 2; emphasis added). She concludes that "instead of establishing rules about how to fish, the rules focus on the outcome and the more detailed implementation decisions would be left to the industry" (Damanaki 2011d, 11–12). While the overall goal of MSY is established at a European level, the method of achieving it becomes an open-ended process that aspires to work with and through the actions of fishermen rather than trying to control them directly. This way of formulating the problem of overfishing has interesting parallels with the emergence of liberal forms of analysis and governmental practice in the late eighteenth century, when policy makers sought to respond to a similar kind of problem: grain scarcity.

In a lecture given at the College de France in 1978, Michel Foucault examines debates over the problem of grain scarcity that arose in the late eighteenth century. He uses this example to illustrate the emergence of a new type of power, biopower, which had as its target the overall health of the population—a concept closely associated with Thomas Malthus, the father of demography. Malthus and other economists at this time were challenging the view that identified the population as a quantifiable number of people inhabiting a particular territory – an assumption that had given rise to policies promoting population growth in the expectation this would lead to an increase in productivity and wealth. The early liberal economists complicated this assumption by arguing that the optimal population in any given territory depended on a whole host of other variables, including the availability of resources, the social and technical capacity to harness them, and levels of consumption. They showed that population was a dynamic, not stable, phenomenon, and that it depended on a series of other factors. From this perspective, policy makers should never try to answer questions about what was best for the population in abstract terms; they could not advance effective policies without proper knowledge of the specific conditions that pertained and the likely effects of their policy decisions. The knowledge relating to demography, Malthus writes, can never be exact because it addresses at its core "the agency of so variable a being as man, and the qualities of so variable a compound as the soil" (quoted in Mayhew 2014, 116). New forms of analysis and governmental intervention now targeted the "population," which in Malthusian terms was not a collection of legal subjects but a "set of processes to be managed at the level and on the basis of what is natural in these processes" (Foucault 2008, 70). But these processes were neither immediately knowable nor amenable to political authority. As Foucault writes, "If one says to a population 'do this,' there is not only no guarantee that it will do it, but also there is quite simply no guarantee that it can do it" (Foucault 2004, 71). The very "nature" of the population thus becomes a *limit* on power, something that must be acknowledged and obeyed if government is to ensure the ongoing health of the population.

In the first part of this chapter, I draw some clear parallels between these historic developments in liberal thought and modes of government and the way overfishing has been problematized over the past fifteen years. Just as early liberal thinkers called for more precise knowledge about social and natural phenomena so they could be governed more effectively, so too have fisheries managers, environmentalists, and scientists advised the European Commission and member states to manage the fisheries according to the real biological and economic processes that have resulted in overfishing. By *accepting* the problem of overfishing as a real phenomenon, the focus of government shifts toward understanding the underlying causes of overfishing in order to redirect them toward the overall goal of MSY. This generates a new liberal governmentality that appears to "liberate" the underlying economic behavior of fishermen and the unpredictable behavior of fish stocks. By analyzing the reasoning behind two recent and significant policy changes in the European CFP—the ban on discarding and the introduction of transferable quotas—I show how these policies are rooted in the identification and shaping of the individual freedom of fishermen. In the second part of the chapter, I examine how this freedom is enabled within certain limits. These limits are determined not by a particular authority (the EU or the Irish government) but by the assumed "natural" thresholds of fish stocks and the dynamics of the market. According to this liberal calculation of freedom, actions and activities that work against bioeconomic "nature" are not just unproductive but irrational.

The Liberal Problematization of Scarcity

Established in 1983, the European CFP was initially a response to conflicts breaking out among member states over access to fish stocks (Sissenwine and Symes 2007). Prior to the introduction of the CFP, European Economic Community (EEC) member countries pursued their own objectives when it came to the fisheries. In most countries this amounted to an intensive capacity-building program that ultimately created conflicts at sea over fish stocks. The growth of competition among European countries eventually led to talks aimed at devising a system for dividing fish stocks equitably on the basis of the historical catches of each national fleet. Based on the recommendations of its own scientific advisers, the Scientific, Technical and Economic Committee of Fisheries (STECF), and ICES, the European Council awarded each member state a proportion of the Total Allowable Catch (TAC). The national share of the European catch is then divided between regional fisheries organizations operating within the country. Until recently, the enforcement of these quotas by individual member states formed the basis of fishing regulation. European policy required each member state to monitor the landings of its fleet in order to ensure they complied with the allocated quota. By modernizing and expanding during the 1980s and 1990s, however, fishing fleets often exceeded their allocated quota and depleted fish stocks.[2]

The reform of the CFP in 2002 reflected growing concerns about the state of the fish stocks and the growing overcapacity of the fisheries. According to estimates at the time, yields were exceeding sustainable levels by 40 percent. Having identified these problems, however, the CFP failed to bring about any meaningful policy changes. The main policy tools it resorted to were increased finance for decommissioning (compensating fishermen to leave the fishery) and stricter enforcement of quotas.[3] Neither of these policies were effective (Symes 2005, 2009): scientific knowledge about the status of key commercial fish stocks was worse than it was twenty-five years before; fish stocks continued suffering while fishermen broke the law; and relations between the fishing industry, scientists, and the government deteriorated (see Dietz et al. 2003; Perry et al. 2010; Österblom 2011).

In 2007 David Symes coauthored a report with Erik Sissenwine entitled "Reflections on the Common Fisheries Policy: Report to the General Directorate for Fisheries and Maritime Affairs of the European Commission." Sponsored by the European Commission "as part of an internal process of reflection," the report targets the structural weaknesses of the existing CFP, its fundamentally flawed analysis of the problem of overfishing, and, by extension, the inappropriate management system for dealing with it. The authors considered the CFP a complete failure even on its own terms; over the past twenty-five years, their report states, the EU had presided "over an unparalleled period of decline for Europe's fishing industries" (Sissenwine and Symes 2007, 49). It is worth looking more closely at what the authors mean by a "fundamentally flawed analysis" and what recommendations they make.

The report emphasizes the weakness of the existing quota system, which relies on annual thresholds being set on the basis of scientific knowledge about the state of the fish stocks. To begin with, they argue, this management strategy ignores the inaccuracy of existing scientific data. One of the primary reasons for the inaccuracy of the data is the quota system itself. When fishing vessels exceed the fixed quotas for certain species, their crews must discard fish at sea in order to avoid penalties. Yet scientists base their conclusions on the number of fish landed on shore. Because the proportion of fish discarded in any fishery is estimated to be between 20 and 60 percent, the volume of fish landed does not accurately reflect the total number of fish being taken from the fishery. As a result, the status of roughly 57 percent of European commercial fish stocks is "unknown" owing to the unreliability of catch data. In effect, the escalating levels of discarding over the past two decades has led to a corresponding decrease in the accuracy of scientific knowledge about fish stocks in European waters. A management system based on rather uncertain biological evidence, they argue, does not provide a sound basis for structural long-term stock management.

Even if biological knowledge of fish stocks were accurate, the report goes on, the political "horse-trading" that takes place between members states at the European level ensures that the scientific thresholds provided by ICES each December are not the final quotas agreed on. On top of this, the more the EU and member states try to police these fixed quotas, the more a gap opens up between policy aspirations and their effects. As Sissenwine and Symes put it, "There is a sense in which policy failure actually occurs during the transition from legislative proposal to practical implementation—'a kind of Sargasso Sea, where policy initiatives lose momentum, founder and disappear from sight'" (Sissenwine and Symes 2007, 49). The emphasis on policing inflexible quotas has given rise to a bureaucratic, top-down control regime that they describe as "archaic." They write: "It is clear that the CFP suffers a serious image problem. The Commission is seen as 'regulator and enforcer' rather than as 'facilitator or enabler,' authoritarian and elitist in its unquestioning adherence to conventional fisheries science (stock assessments) and remote, unresponsive and bureaucratic in its relation with the industry" (Sissenwine and Symes 2007, 50). Referring to the unrealistic ambitions of a policy that attempts to regulate the European fisheries directly, they argue that "there is something faintly ludicrous about the Directorate General of fisheries—a bureaucracy probably no bigger than the planning department of an average sized local authority—attempting to regulate the fisheries of an area that stretches through 40 degrees of latitude from the Gulf of Bothnia to the Canary Islands and 60 degrees of longitude from the Azores to the eastern Mediterranean. Even Napoleon might have baulked at an empire of these dimensions!" (Sissenwine and Symes 2007, 63).

Sissenwine and Symes conclude the report by calling for a fundamental reorientation of the CFP. This should begin by acknowledging the existing status of the fishing industry, the fish stocks, and the quality of scientific knowledge as well as the underlying causes of overfishing. They stress that certain outcomes will be inescapable between 2012 and 2020, namely "greatly altered or reduced fishing opportunities constrained by environmental change." Taking these constraints as a starting point, the report asks how fisheries managers, scientists, fishermen, and policy makers can implement a *realistic* system of management:

> There is a sense in which failure is inherent in a policy which attempts to define and subsequently police some form of "sustainable limits." *Wherever those limits are drawn there is an inexorable tendency for the industry to cross the line in pursuit of profit, survival or protection of assets.* The failure of the CFP, therefore, lies in characterising the problem of depleted resources as a series of contingent crises *rather than seeking to understand why unsustainable actions become the norm.* As a result, policy is directed at treating the symptoms rather than the underlying cause: the tendency of capital to innovate, invest and build capacity. (Sissenwine and Symes 2007, 53; emphasis added)

If sustainable development of the fisheries is to move from a vague policy ambition to a practical reality, Symes and Sissenwine argue, unsustainable practices and events have to be addressed as the real outcomes of broader social, economic, and ecological processes. Policy makers must therefore recognize and accept that the causes of overfishing consist of *real* economic motivations and behaviors which interact with the biological patterns of different fish stocks.

These realities are not going to disappear by simply increasing regulation and policing. The authors stress the need to recognize how the crisis of overfishing results from a mode of production that tends toward the exhaustion of marine resources. Rather than ignoring this tendency or trying to police it at a growing expense, policy makers must *accept* it and understand its underlying dynamics so that policy can be more effective at channeling these causes toward desirable outcomes. The report advises the EU to stop focusing on the enforcement of "sustainable" fisheries based on questionable scientific data, and instead concentrate on measurable goals such as the sustainable yields of fish stocks, limiting or banning discards, and the introduction of rights-based access to the fisheries. The authors insist that such goals can only be achieved through an ecosystems-based approach to fisheries management. In short, this approach strives to achieve the goal of sustainable fisheries by accepting the uncertainty and complexity of the human and nonhuman interactions within any ecosystem. This includes the behavior of fishermen ("to innovate, invest and build capacity") and the different biological characteristics of fish species and stocks.

There are remarkable similarities between the arguments contained in the Sissenwine and Symes report and those identified by Foucault in Louis-Paul Abeille's *Lettre d'un négociant sur la nature du commerce des grains*, a text from 1760s France. Abeille was a student of the Physiocrats, a group of political theorists who addressed the question of grain scarcity and, more specifically, the failure of existing government policies to effectively respond to it in the mid to late eighteenth century. At that time grain scarcity was a serious issue for French and British authorities. This was not simply because the population was going hungry but because lack of grain in the towns was precipitating rioting and social disorder.

According to Foucault, the dominant response to problems of grain scarcity before the 1750s was a policy of "anti-scarcity." In other words, the authorities tried to prevent grain scarcity by directly intervening in the grain markets, prohibiting the withholding of grain by forcing peasant producers to supply the market even if they made a loss. In Britain, common, customary, and statute laws prohibited the activities of middlemen and profit seekers who benefited from the production and sale of grain at the expense of others.[4] Despite these strong, prescriptive policies, the problem of grain scarcity continued. This failure on the part of the authorities subsequently provoked a broad debate among political economists and thinkers on the rationale behind these interventions and the nature of the problem itself.

In his text, Abeille begins by questioning the basis of "anti-scarcity" policies. The specter of scarcity is nothing but an "imagined" event, a "chimaera," as he calls it. The problem, Abeille argues, is that scarcity is framed as something "bad" or "unnatural" that must be stopped. What are the results of this approach? First, the state wastes considerable resources on the prevention of scarcity through constant vigilance, discipline, and regulation by an administration that is both costly and inefficient. Second, and more importantly, this policy often meant ruin for peasants, who were essentially punished in order to ensure low grain prices, leading to adverse effects in terms of subsequent grain production. With fewer profits, peasants had less money to buy and then sow grain or to invest in more extensive cultivation. In the event of drought or poor weather conditions, this constantly uncertain level of grain production would mean the return of scarcity, a situation that the existing government could do little about. In the final analysis, the policies of the authorities were always going to fail if their sole objective was to prevent scarcity. For Abeille, the roots of ineffective government were policies derived from what was thought desirable rather than from a careful analysis of what was actually happening. Rather than trying to prevent scarcity, therefore, policy makers should begin by accepting the "reality" of scarcity, its existence as a phenomenon that cannot be avoided and thus should not be repressed.

The consequences of this analysis are that scarcity itself is no longer the primary object of analysis or intervention. Rather, such events are understood to be the result of a much wider set of social, economic, and environmental factors. In effect, the object of analysis becomes everything other than the event of scarcity. Abeille goes on to identify these.

First, there is the biophysical reality of the grain itself, with "everything that may happen to it and will happen to it naturally, as it were, according to a mechanism and laws in which the quality of the land, the care with which it is cultivated, the climatic conditions of dryness, heat, humidity, and finally the abundance or scarcity, of course, and its marketing and so forth, also play a part" (Foucault 2004, 36). Second, there is the economic motivation of the grain producer. Abeille argues that "instead of subjecting them [peasants] to obligatory rules, we will try to identify, understand, and know how and why they act, what calculation they make when, faced with a price rise, they hold back grain, and what calculation they make when, on the other hand, they know there is freedom, when they do not know how much grain will arrive" (Foucault 2004, 40). Finally, there is the broader socioeconomic context with its fluctuations and dynamics unfolding at regional, national, and global levels. Beginning with these different elements and the way they interact through the market entails a radical shift in the knowledge and practice of government. It means "trying to grasp them (economic behavior, the production of grain) at the level of their nature, or let's say—this word not having the meaning we now give it—grasping them at the level of their *effective reality*" (Foucault 2004, 47; emphasis added).

Abeille asks what would happen, for example, if the government stopped maintaining an artificially low price for grain (to ensure it was affordable) and at the same time removed the ban on hoarding and on exports? To begin with, if prices were high because grain was scarce in one market, the price would attract other grain sellers, thus alleviating the scarcity. By encouraging peasants to profit from the production of grain, the market would also encourage them to extend cultivation as they sought to make more money at each harvest, providing them with the necessary money to buy more seeds for sowing and to rent land. Even if climatic conditions were adverse, the increase in cultivation would mean better harvests. By allowing the market to function in this way, the probability of scarcity can be reduced:

> Instead of imposing a battery of restrictions on the grain trade, the preferred policy is to allow the free flow of commodities and achieve the self-regulation of prices, allowing profits to be made which will then be invested in new cultivation, increasing the amount of grain for sale the following year, and hence lowering prices, just as allowing the possibility of imports will discourage hoarding more successfully than prohibitions. Of course, this will not eliminate revolts entirely, but it will deprive them of their justification because in acting in this way the sovereign will be acting in conformity with "the nature of things" and not by means of prohibitions for whose ineffectiveness he will be held responsible. (Donzelot 2008, 118)

By inserting the phenomenon of scarcity into a general field of knowledge, the causes, conditions, and effects of that phenomenon are brought into new relations with one another. Foucault calls this new field of knowledge political economy. Political economy describes an extensive and continuously productive field of knowledge, "an economics, or a political-economic analysis, that integrates the moment of production, the world market, and, finally, the economic behaviour of the population, of producers and consumers" (Foucault 2004, 41).

The total sum of interventions that arise from this expansive and dynamic analysis (removing tariffs, investing in infrastructure for markets, educating grain producers, etc.) is what Foucault calls an "apparatus of security," a way of "arranging things so that, by connecting up with the very reality of these fluctuations, and by establishing a series of connections with other elements of reality, the phenomenon [scarcity] is gradually compensated for, checked, finally limited and, in the final degree, cancelled out, without it being prevented or losing any of its reality" (Foucault 2004, 37). This "apparatus" describes the totality of interventions and policies that shape a particular sphere of life, such as grain production, compensating for and balancing out different elements as they emerge. From year to year, for example, the same policies may not hold as economic and environmental conditions change, resulting in additional measures to compensate for the changes. The "nature" of this economy is "not something on which, above which, or against which the sovereign must impose just laws" (Foucault 2004, 75). The task of the sovereign is to graft procedures of government onto the economy and thereby shape it toward general goals, avoiding the event of scarcity. As Tiziana Terranova writes, governmentality does not "aim to suspend the 'interplay of reality' that supposedly belongs to the domain of nature, but *operate within it*" (2008, 239–40; emphasis added).

Just as Abeille and other early political economists formulated the problem of grain scarcity, so are analysts and policy-makers formulating the problem of overfishing today. The recent ban on discarding and the introduction of a more flexible quota system provide a good example of how this liberal economic analysis translates into a series of interventions that are coordinated around the economic actions of individual fishermen and the unpredictable behavior of different fish stocks.

Discarding refers to the practice of throwing unwanted fish overboard at sea. The unpredictability of the sea and the mixed swimming patterns of many species of commercial fish mean that there will always be a certain quantity of unwanted fish in any catch.[5] This is called the "by-catch." However, the quantity of discarded fish has increased dramatically over the past thirty years as fishing fleets have expanded their fishing capacity and target only the most profitable species. As early as 1975, the FAO estimated that one-third of all marine resources harvested were wasted in the post-harvest process. While some fish were lost during the wider distribution process, the bulk of the waste occurred at the point of capture, through the deliberate discarding of fish at sea. In the European fisheries, the problem of discards has now become acute. While the figures are inaccurate, it is clear that in many fisheries discards occur at high levels, with mixed fisheries estimated to have a discard rate of anywhere between 20 and 60 percent (Commission for Environmental Cooperation 2011a).

In recent years, the problem of discarding has received considerable attention from environmental NGOs, policy makers, and the fishing industry. Discards came to represent the destructive logic of industrial-scale fishing, the irrationality of commercial fishing as a whole, and a rallying point for all those concerned about the future of fish stocks. Alongside emotive documentaries like *The End of the Line*, Hugh Fearnley-Whittingstall, a celebrity chef in the UK, led a high-profile media campaign to ban the practice of discarding in the European fisheries. "Hugh's Fish Fight" presented itself as a "popular movement" to ban an "unethical" environmental practice. Enlisting social media and producing well-publicized documentaries, it was a very successful campaign that garnered widespread television and radio coverage. It collected nearly one million signatures in a bid to have the European Commission introduce a ban on discards. The EU Commissioner for fisheries, Maria Damanaki, invited the campaigners to Brussels and stated: "I consider discarding of fish unethical, a waste of natural resources and a waste of fishermen's effort. But I would like to go further—since our stocks are declining, these figures are not justifiable anymore" (Damanaki 2011a).

The EU has now decided to gradually implement a ban on all discards by 2019.[6] Under the ban, all quota fish are to be landed and recorded. While the fishing industry has voiced some reservations about how this is going to be implemented, the consensus is that the ban makes economic, biological, and ethical sense. It will not only prevent the needless waste of valuable fish but also improve the scientific data required to assess and manage stocks. The decision to ban the practice of discarding was unsurprisingly one of the headline stories of the new CFP and one of the main reasons it received such widespread praise. Yet the apparent simplicity of this populist response obscures the reasoning behind the ban and its far-reaching consequences for fishermen and fisheries management.

The justification for the ban was that it would reverse incentives for fishermen to discard fish in a "meaningful" way. While previous administrations had encouraged technical conservation measures and more selective fishing, these had been voluntary and generally represented a significant burden for fishermen: a fisherman already struggling to cover the rising costs of diesel, for example, has difficulty affording a new fishing net with a larger mesh size to avoid catching juvenile fish. Similarly, it takes time, considerable effort, and short-term sacrifices for fishermen to come together and make collective decisions to avoid certain fishing grounds at certain times of year, or to work with scientists in order to record and monitor the breeding patterns of particular fish stocks. As long as discarding was allowed, fishermen could make calculated decisions about their catch *after* the point of capture and thus not have to worry too much about adopting more efficient but costly fishing equipment and new techniques. The ban on discards effectively takes this capacity for decision-making away from fishermen, shifting it instead to the point of capture itself: "It [the discard ban] will also be a driver to avoid unwanted catches and will deliver a level playing field to change the fishing strategies of fishermen" (Commission for Environmental Cooperation 2011c, 1). Fishermen will not want to catch noncommercial or undersized fish because they will not make any money from what they land. Alongside this economic incentive, the EU has stated that under the discard ban, if fishermen land commercial fish stocks in excess of their available quota, then they will be held responsible and remain subject to the normal penalties for overfishing. In addition to the threat of individual penalties, there is the possibility that an entire fishing ground could be closed if fishermen continue to land a particular species of fish beyond the allocated quota. If the quota for cod in an area is exhausted, for example, then those who are fishing for plaice or haddock—fish that are often caught with cod—would also be prevented from fishing those waters. This restriction puts considerable pressure on fishermen to adopt more selective fishing practices and to collaborate with one another and with scientists in order to avoid the almost certain loss of fishing opportunities.

Although the discard ban will force fishermen to become more selective in their fishing practices, they will not be able to ensure that the number of fish they catch (and land) perfectly matches their allocated quota. Improved knowledge of fishing behavior, distribution, and swimming patterns as well as the design of more selective gear and fishing practices will account for *some* of the uncertainty inherent to fishing, but it will not eliminate it. In response to this situation, the recent CFP (2013) has also stated that some form of transferable quota system represents a necessary policy response to the mismatch between available quota and the quantities of fish that continue to be caught and now landed. Effectively, the introduction of an individual transferable quota system would allow individual fishermen to buy, sell, and lease quotas from their Producer Organization or other fishermen. If a fisherman lands fish over his quota, he would be able to buy or lease more quota from another fisherman. If a fisherman does not fill his quota, he will be able to sell the remainder to someone else. The presumed benefit of this policy is that it will compensate for the uncertainty of fish catches, ensuring that the fish landed and the available quotas are better matched. As with the ban on discards, the policy will operate by incentivizing fishermen to be more selective in their fishing (to avoid having to buy more quota) and by allowing less viable fishermen to exit the fishery and sell their quota to more efficient operators. Over time, this "natural" restructuring is expected to generate a more efficient fishery that ensures a match between what is caught and the quotas set according to the best scientific data. Matching these two figures is crucial to achieving the goal of MSY.

While this flexible quota system is not being enforced by the EU, member states are free to decide whether they want to implement some form of tradeable or transferable quota system within its fishing fleet. This reflects the ongoing debate around tradeable quota systems and the negative social, economic, and environmental effects they can have on the fisheries (Crean and Symes 1995). The fixed quota system has been the bedrock of the CFP since it was established in 1983. It is based on the principle of "relative stability," ensuring that each member state is *guaranteed* a fixed share of the Total Allowable Catch (TAC) set each December by ICES for all European fish stocks. Each member state then distributes its allocated quota to regional Producer Organizations (PO), which in turn distribute the quota between specific fishing vessels according to their size. While these quotas are issued on an individual basis, *they are nontransferable*. The amount of quota may rise or fall, but the *proportion* of the overall quota granted to regional fisheries remains the same. The "relative stability" principle ensures rights of access to fisheries' resources for fishing communities, especially those that are dependent on fishing as their main source of income and economic activity.

The arguments for and against the introduction of individual transferable quota continue.[7] But the decision to implement the policy now should be understood within the context of recent analyses and reflections on the problem of discarding and overfishing within the European fisheries.[8] In analyzing the operations, effects, conditions, and transactions of bioeconomic "nature" (understood as the interplay of environmental factors, economic self-interest, and the market in producing phenomena such as overfishing), a certain rationality emerges that evaluates the efficacy or inefficacy of certain decisions or policies on the basis of their likely effects and how they might contribute to the overall goal of MSY—a "virtuous cycle of fishing." Instead of asking what authorizes the government to introduce transferable quotas, this kind of liberal reasoning asks what will happen if, at a given moment, we introduce this policy: what will it mean in terms of limiting discarding, improving scientific knowledge, bringing the fisheries towards MSY, and so on. Distinctions between "right" or "wrong" are thus no longer meaningful when "truth" is understood as efficacy and the success or failure of a particular action is predicated on the overall balance of available resources to resource users.[9]

This liberal formulation of scarcity and the emergence of political economy as the intellectual instrument of government is, Foucault writes, "a crucial moment since it establishes, in its most important features, not the reign of truth in politics, but a *particular regime of truth* which is a characteristic feature of what could be called the age of politics and the basic apparatus of which is in fact still the same today" (Foucault 2008, 17–18; emphasis added). The distinction between a "reign of truth" and "regime of truth" is helpful in understanding the novelty and power of liberal forms of government, specifically the production of new forms of liberal freedom alongside new forms of illiberal exclusion in the "sustainable" development of the fisheries.

Homo Economicus and the New Nature

Hugh Fearnley-Whittingstall's "Fish Fight" campaign to end the practice of discarding targeted the "EU's crazy laws." It identified the "overregulation" of the European CFP as the main reason fishermen were having to dump fish overboard. Similarly, fishermen in Ireland and across Europe protested against what they saw as an inflexible quota system that effectively incentivized the discarding of good quality fish: in 2012 fishermen in Ireland landed fish they were supposed to have dumped at sea and then handed them out for free to members of the public (Furlong 2012). In one sense, the discard ban and the implementation of ITQ respond to these demands. Reformers identified the regulatory and economic incentives for discarding as "obstacles" to the rational exploitation of marine resources by fishermen; the enforcement of strict quotas by the EU effectively *forced* fishermen to discard. The ban on discards thus represents the "liberation" of fishermen from an "irrational" system. Similarly, the implementation of ITQ marks a further liberation of fishermen from the "inflexible," fixed quota system that effectively forced fishermen to discard nonquota species and failed to take into account the unpredictable swimming patterns of fish. The flexibility of the new quota system will restore the "natural" interplay of the market to ensure coordination of the available quotas and fishing effort. Rather than being "imposed" on the fisheries, these measures are thus seen to reflect the "real" economic and biological dynamics operating within the fisheries.

For Abeille and the Physiocrats, the hallmark of good government is *allowing* the market to function and letting the laws of supply and demand dictate where and how much grain is grown, at what time, for whom, and so on. This is a reversal of how the state had previously interacted with the economy: attempting to control prices and limit circulation. In this liberal account, the "free" market is understood as the sphere of exchange and circulation, ensuring the free movement of people and goods from places of abundance to places of dearth. When there is high demand for something (labor, grain), the promise of financial reward will ensure that resources will flow there. The market also ensures that natural resources are cultivated to their maximum productivity through individual economic incentives. Unlike the previous "anti-scarcity" mode of regulation, one based on discipline and prescription, the liberal mode of governing seeks to remove any obstacles to the "natural" functioning of the market. As Foucault concludes, "The game of liberalism—not interfering, allowing free movement, letting things follow their course; *laisser faire, passer et aller*—basically and fundamentally means acting so that reality develops, goes its way, and follows its own course according to the laws, principles and mechanisms of reality itself" (Foucault 2004, 48).

This approach all rests on the assumption that individuals will act in their own self-interest and that they will calculate the best strategies for cultivating grain, how much to cultivate, where to sell it, and when. By allowing (and assisting) individuals to make these decisions in response to changing environmental and market conditions, governments can achieve overall goals that are for the common good. The entire edifice of liberal political economy and good government is thus grounded in the rational, economic subject: *homo economicus*. This individual responds rationally to modifications in the environment, such as rises in price, periods of drought, or new agricultural technologies, in ways that government programs can define and measure. *Homo economicus* is the node through which government of the population passes: "*Homo economicus* is the one island of rationality possible within an economic process whose uncontrollable nature does not challenge, but instead founds the rationality of the atomistic behaviour of *homo economicus*" (Foucault 2008, 282). All will be well as long as "every individual functions well as a member, as an element of the thing we want to manage in the best way possible, namely the population" (Foucault 2008, 43).

While liberal governmentality relies on the freedom of individuals to act in their own economic self-interest, its only concern with this freedom is that it can help achieve generally defined goals—in this case, the balancing of fishing effort with available fish stocks.

In other words, the fate of particular fishermen is not the government's concern, only the extent to which their behavior is calculable and thus manageable within the overarching framework of bioeconomic sustainability. Governmentality—the "art of governing"—thus consists of a fluid combination of regulations, policies, and interventions that together compose a reality in which certain actions (such as reducing by-catch) become rational and others become irrational.

Dr. Sam Lee is a marine scientist in the Irish Marine Institute (the national body responsible for undertaking, coordinating, and promoting marine research in Ireland) and is responsible for leading "industry-science" partnerships.[10] The motivation for these partnerships is to bring together industry representatives, scientific advisors, and fisheries managers to discuss the state of the fisheries, or as Lee described it, "to ensure that what I'm saying from an advisory point of view is reflected with what these guys are seeing on the ground. *So the science matches the truth*" (Lee 2009; emphasis added). These partnerships have been part of European policy since 2002, but Lee told me that they had not been successful in the beginning because fishermen had "no reason to come to the meetings." In the last few years, however, the situation has changed as a result of increased pressures on fishermen to collaborate with fisheries managers and scientists.

Two weeks before I met him, Dr. Lee and twenty skippers had attended an industry-science meeting in Dunmanway, County Cork. Lee told me later that he had been expecting a really tough meeting, but it turned out to be one of the most productive he had ever experienced: "it was just like forget about what happened in the past. We have to get the science up and running, what do we need to do?" They spent six or seven hours going through the problems "as they saw them, and the problems as the Marine Institute saw them." Each party identified an issue and asked the others how they might collectively go about resolving it. The fishermen proposed keeping private diaries on board boats so there would be a "constant record of what happened." Another suggestion was to install CCTV cameras on board. In addition to this meeting, Lee described recent efforts by other groups of fishermen to compensate for shrinking scientific budgets and inaccurate assessments by measuring stocks themselves. They conducted surveys to prove that the stock of Celtic Sea cod was far bigger than what scientists had claimed. The fishermen decided collectively to take a cut in their quota in order to pay for one of their boats to work full time at collecting samples. According to Lee, this initiative was the first of its kind in Europe: "it has always been the state that pays: we do the job, there's the data . . . they [fishermen] are so concerned about the Celtic sea cod, that there is no analytic data, that they really are putting their money where their mouths are" (2009).

Lee suggested there had been a fundamental change in "culture": "fishermen are now realizing that historically, just by not giving us data it was a kind of protest, you're not allowed in here, you're not allowed onto our boats, is actually coming back to haunt them . . . there is now a bit of a scramble with everyone now suddenly loving the scientist." He used the term "levers" to describe the various pressures that are making fishermen change their attitude toward scientists. Dave McCarthy, a fisheries manager, corroborated this view. He told me that in the space of two years, the situation had changed dramatically: "things have got more critical and decisions have become more important," he said. "I think the pace of change has increased. Fishermen are facing more serious decisions than they have done in the past. . . . There is less room to maneuver" (McCarthy 2009).

One of the most significant "levers" in this regard is the "precautionary principle." While this contentious management tool has been part of European environmental policy since 2000, its first application in the fisheries was in 2012. The precautionary approach holds that in situations where scientific data is insufficient to develop an effective management plan, the precautionary principle requires "proportional" action to prevent a predicted outcome, such as the collapse of a fish stock. It is a form of risk management: to stop an event before it happens, to predict, without sufficient information, the likelihood of something happening. In the fisheries, where the status of 50 percent of commercial fish stocks is unknown, the applicability of such a measure is obvious. It was used in 2012 in an area off the coast of Scotland and Northern Ireland. Overnight the total allowable catch and allocation of days at sea for fishermen were cut by 25 percent. As Dr. Lee explained, these cuts essentially meant that a quarter of the fleet was prevented from making a living, a setback from which they are unlikely to recover.

Between 2003 and 2005, the EU also established a program to target misreporting of fish catches. This included the threat and, in some cases, imposition of large sanctions on member states that were found to be ineffectively policing their fisheries. Heavy fines imposed on Ireland in 2006 resulted in the government establishing a new division of the fisheries department, the Sea Fisheries Protection Authority (SFPA). Overnight the number of fisheries officers in Castletownbere, the port where I conducted my fieldwork, went from one to thirteen. A new building housed them on the pier. Suddenly, fishermen who had never had their boats checked before were having them searched every time they returned to port. The new fisheries act (2006) that established the SFPA also led to the criminalization of infringements of fishing quotas and unreported catches. Incorrect entries in logbooks, for example, now resulted in a criminal charge and possible jail sentence, a massive worry for many fishermen who visited children living in America. I was told that some older fishermen who were fined large amounts for relatively small infringements found the controls so stressful that they stopped fishing altogether. A man who had fished for forty years was fined twenty-five thousand euros after exceeding the quota for monkfish by 400 kilograms. He gave up fishing immediately because he could not deal with the fear of coming into harbor and being "harassed." In this context someone painted the words "Scum out" on the side of the SFPA building in Castletownbere.[11] Even a local fisheries officer, Gerry Owens, thought the Irish government's response to the EU sanctions was extraordinary:

> These guys [fishermen] are under huge pressure. You've got fellows here being raided by fraud squad. Huge, huge governmental pressure being applied on fishermen the likes of which you wouldn't see

> against drug gangs. . . . At the end of the day you wonder where it's all going. Is it being driven by EU, in a not overt way, to slowly apply more and more pressure to manage an exit from the fishery for a lot of these fellows? (Owens 2009)

Understood from a liberal perspective, previous acts of noncompliance were "rational" when fishermen were advancing their own economic self-interest against stifling forms of regulation, such as the fixed-quota system, which fishermen often ignored. The ban on discards and the introduction of ITQs, by contrast, are understood to *accommodate* the economic self-interest of individual fishermen, as well as the unpredictable nature of fishing and fish stocks. In this account, fishermen are enabled to pursue their own self-interest in order that the common goal of balanced fisheries can be achieved. The flip side of this is that violators are not simply breaking the law but acting "irrationally" against the common good. What starts to become visible is the janus-face of liberal governmentality: on one side, the promise of individual freedom, and on the other an economic rationality that is all the more dominant because it claims to act in accordance with what is "natural" or real.As Foucault writes,

> You can see therefore that the principle of laissez-faire in the physiocrats, the principle of the necessary freedom of economic agents can coincide with the existence of a sovereign who is all the more despotic and unrestrained by traditions, customs, rules, and fundamental laws as his only law is that of evidence, of a well-formed, well-constructed knowledge which he will share with the economic agents. (Foucault 2008, 285)

This aspect of liberal reasoning is captured best in the public statements of EU Commissioner Maria Damanaki. In a speech given at a fisheries committee meeting in 2010, Damanaki accepted the perception of transferable quotas as a "capitalist response" to the crisis of overcapacity in the fisheries but argued that they were necessary to resolve a problem "*which had no other obvious solution*" (2010; emphasis added). "If I were a banker," she went on, "I would say our fish stocks are underperforming assets. Instead, I want a capital of healthier fish stocks giving rich interests, in the form of landings, to our fishing industry. I want to maximize the economic return to fishing communities. . . . Only under this precondition can fishermen continue to fish and earn a decent living out of their activities" (Damanaki 2011e). She argued that an ITQ system, done properly with appropriate and meaningful safeguards, would yield obvious benefits and few if any negative side effects. She cited the example of Denmark, where ITQ reduced the pelagic fleet by 50 percent in three years and the demersal fleet by 30 percent since 2007 (Damanaki 2010).[12]

Commissioner Damanaki has argued that ITQ should be seen less as property rights and more as user rights that grant access to a common resource for those who are able to fish responsibly.[13] She reinforces an argument that has been made since at least 2009, when the reform of the Common Fisheries Policy was launched: "so far the fishing industry has been given free access to a public resource and management costs have been largely incurred by taxpayers. . . . *Those who exercise responsibility in a proper and effective manner should be the ones to enjoy the access to fish stocks*" (Commission for Environmental Cooperation 2009, 12; emphasis added).[14] By redefining the European fisheries as a "common" resource, the CFP transforms the right of any individual or group to exploit it into a privilege that must be earned. Fishermen must now show they are capable of exploiting this "common" resource for the indefinite future ("the good of all present and future generations").[15] The force of this argument is subsequently articulated through the escalating surveillance and policing of the fisheries, culminating in Commissioner Damanaki's recent commitment to ensure a "zero tolerance" policy toward illegal fishing (Commission for Environmental Cooperation 2011b): "we can no longer allow even a small minority of fishermen to ignore the rules, and get away with it. Apart from being unfair this also undermines conservation efforts; it disrupts markets with unfair competition; it penalises law-abiding fishermen and chokes the circle of compliance; *and, most importantly, it destroys fish stocks*" (Damanaki 2011d; emphasis added).[16]

In Commissioner Damanaki's statements, we can identify a new and productive relationship between the freedom (and responsibility) of the individual fisherman to conduct themselves "rationally" and the achievement of the overall goal of MSY. Those who are unable or unwilling to fish within the "natural" limits of the fish stocks can thus be penalized and banned on the grounds of protecting the common good (Opitz 2011; Rabinow and Rose 2006):

> Whereas the subject of liberalism has to follow his interests by taking reasonable risks, the deconstituted, dangerous subject is portrayed as a deeply uneconomic subject. . . . Confronted with the dangerous subject, governmentality encounters an interest that consumes the rational subject entirely—and turns it into an irrational, unintelligible, destructive agent outside the bounds of humanity. (Opitz 2011, 110)

The distinction between the liberal subject, *homo economicus*, and the illiberal subject whose activities work against the common good helps illustrate what is at stake in the new model of fisheries governance. The "rational" conduct of fishermen in response to new regulations is not simply a question of profitability within a narrow sphere of economic production but also part of a broader *biopolitical* rationality constituted around the common goal of protecting the health of fish stocks. This biopolitical rationality gives rise to new forms of knowledge (calculating the behavior, characteristics, and tendencies of fishermen and the fisheries they exploit) and a new mode of government, an "apparatus of security" that works on these "natural" processes, shaping the conduct of individual fishermen through a series of incentives and disincentives.

Rethinking Enclosure

As the crisis of overfishing has worsened, attention has focused on the inadequacies of existing forms of fisheries management that have sought to directly regulate the activity of fishermen according to fixed quotas. The new approach to fisheries management strives above all to better understand, account for, and allow the free interplay of fishing activity and fish stocks. During my research, I found that the fishermen, policy makers, environmentalists, and scientists I interviewed were supportive in principle of a ban on discards and the introduction of some form of flexibility in the quota system. This was never called "privatization" but was understood above all as a pragmatic response to particular and urgent problems that were depleting the fish stocks.

At one level, the introduction of measures such as transferable quotas into the European fisheries appears to tell a familiar story: the effective privatization of one of the last remaining "commons" and the extension of neoliberal environmental policies. This account might be true, but it says little about how these policies emerge from a process of critical and widespread reflection on the nature of the problem of overfishing and a series of concrete analyses that seek to map onto and shape certain prevailing bioeconomic tendencies. The resulting analyses and the policies they give rise to do not originate from a prior belief in the "truth" of the market. They originate from a motivation to bring about MSY, a specific, measurable goal that can only be achieved through policies that effectively work on and change the behavior of fishermen. The justification or "truth" of these policies thus lies in the degree to which they can be shown to move the fisheries toward this goal by measurably reducing fishing effort.

In this chapter I have sought to show how these transformations in fisheries management can be understood alongside Foucault's analysis of the liberal response to the problem of grain scarcity in the second half of the eighteenth century. Rather than denying the "reality" of scarcity, writers like Abeille in France advised the sovereign to do all it could to understand the causes of scarcity, from the point of production to distribution and consumption, so these processes could be compensated for and channelled in such a way that the phenomenon of scarcity was permanently displaced and controlled. The result was the emergence of a new "apparatus of security" that was more than a set of policies; it was a widely distributed and dynamic arrangement of bodies, resources, technologies, and laws that shaped individual behavior and their interaction with biophysical environments in order to ensure the productive use of resources and labor. In this new mode of government, "Nature" no longer provides a fixed or stable point of reference. "Paradoxically," Lemke writes, "the liberal recourse to nature makes it possible to leave nature behind or, more precisely, to leave behind a certain concept of nature that conceives of it as eternal, holy, or unchangeable" (2010, 46). Rather than ordering the world to fit a particular vision of how it *should* be, liberal governmentality enlists the "natural (bio-economic) processes of life" as the starting point and *limit* of power:

> For political economy, nature is not an original and reserved region on which the exercise of power should not impinge, on pain of being illegitimate. Nature is something that runs under, through, and in the exercise of governmentality. . . . It is not background, but a permanent correlative. Thus, the *economistes* explain, the movement of population to where wages are highest, for example, is a law of nature; it is a law of nature that customs duty protecting the high price of the means of subsistence will inevitably entail something like dearth. (Foucault 2008, 15–16)

"Nature" in this sense is not a static object or thing but the complex interplay of economic and biophysical agencies. It is always outside any exact knowledge or science and thus always beyond the knowledge claims of any particular authority. According to Foucault, Adam Smith's concept of the "invisible hand" was significant because it alluded to the opaqueness of the market, the fact that it was never knowable in a precise fashion. But this limit is what energizes liberal forms of government and the production of knowledge on which it relies. The recognition that individuals, the economy, and the environment are fundamentally uncertain and complex demands the production of more data and analyses, the constant incorporation of new phenomena as they emerge: "new elements are constantly being integrated: production, psychology, behaviour, the ways of doing things of producers, buyers, consumers, importers, and exporters, and the world market" (Foucault 2004, 45).

The need to expand and transform the forms of knowledge required to manage the problem of scarcity finds a contemporary parallel in the critique of how scientific knowledge was applied in former models of top-down environmental management (Freire and

García-Alut 2000, 376; Paterson et al. 2010). As sociologist Ulrich Beck argued over twenty years ago, science is no longer able to provide the "socially binding definition of truth" required to respond to all manner of new "risky" problems besetting the world (1992: 156). Rather than seeing this uncertainty as a problem, however, other theorists have welcomed a new era of "post-normal" science, with the promise of a more inclusive, deliberative process of knowledge production in response to discrete problems (Francis and Goodman 2009; Funtowicz and Ravetz 1993; Ludwig et al. 1993). Environmental theorist Tim Forsyth writes, "This alternative conception of science [was] willing to surrender claims to universal validity in exchange for knowledge *that nears some local and circumscribed utility*" (2003: 165; emphasis added). The emphasis shifts toward "how we can make good decisions the right way" (Collins 2002; Leach 2008; Leach et al. 2007).

This is the pragmatic "spirit" of critique that informed Sissenwine and Symes's report to the European Commission in 2007. As they conclude, "It is not enough for science to be right. It needs to interface with fishery management decision-making processes in a manner *that helps managers make the 'right' decisions*" (Sissenwine and Symes 2007, 29; emphasis added). Instead of debates over values once the facts have been decided by scientists, the value of marine life/fish stocks provides the starting point for largely localized and technical forms of problem-solving among fishermen, scientists, and fisheries managers (Pellizoni 1999; see also Kooiman et al. 2008). There is no single (scientific) authority prescribing what policy is right or wrong because it is precisely this partisan approach to environmental management that is identified as the problem.[17]

The fisheries scientists and fisheries managers I met during my research constantly emphasized this different approach to fisheries management. I did not meet a single marine scientist who believed that biological data alone could bring about the changes required to achieve MSY. Jill Donovan, a marine biologist working in University College Dublin, told me that most scientists believed they had more than enough hard science to prove fish stocks were declining. The challenge now was how to bring about "real" changes in fishing effort and stock management "on the ground" (Donovan 2010). Another researcher—Bill Murphy, who had a background in marine biology and was working for Bord Iascaigh Mhara (BIM), the Irish fisheries agency—called for similar work:

> I don't think it's anything to do with biology and improving biological knowledge of fish stocks. *I think it's purely a governance issue.* Not purely but much more so. . . . You can know everything there is to know about the biological, and hard science, about how all the dynamics of how things work in the ecosystem but unless we improve the governance we're still going to make a mess of the stocks. So I don't see it as a biological issue. I think there's a recognition of that fact dawning. (Murphy 2009; emphasis added)

In this new open-ended process of data collection, consensus-based decision making, and pragmatic policy making, scientists, fisheries managers, and fishermen are brought together to better coordinate and understand the interplay of biological and economic processes. The production of this knowledge and the way it is used to verify or falsify particular policy interventions create what Foucault calls a "regime of truth."[18] The malleability of this mode of governing allows its proponents to abdicate any stance other than what the "facts" appear to tell us.[19]

The historian Jeanette Neeson makes a similar case in her analysis of the arguments used to justify enclosure and "improvements" in Britain in the late eighteenth century. While arguments for enclosure had been made throughout the eighteenth century, there was a strong counterargument that sought to defend the many commoners and small farmers who would be displaced through the process. The defense of common rights was based on historical claims as well as social arguments that recognized that enclosure would create a large dispossessed population with no way of sustaining themselves (Neeson 1996). However, as the harvest-crisis decades of the 1780s and 1790s unfolded (exacerbating the problem of grain scarcity), the critics of the commons led a growing movement for the cheaper and easier enclosure of wastelands. Importantly, Neeson observes, such measures were justified not only because they would deliver greater profits for the land-owning elites but also because they advanced a more general understanding of productivity that was connected to the well-being of the population. "As they saw it," Neeson writes, "common right stood in the way of modernization. Accordingly, they could not approve of it, and they could not see, *in the larger terms of national interest* how common-right economies could be allowed to survive" (1996: 35; emphasis added). The "larger terms of national interest" refers to the Malthusian concept of population, a general, economic figure that was always primary to the specific needs of communities in the analysis of political economy.

While Neeson is quick to point out that the critics of the commons were often unable to see its material, situated value, she illustrates how the arguments for enclosure developed into a powerful series of justifications that relied on a *new political-economic analysis*: the connection of discrete events such as the increased yields from enclosure, access to markets, and the value of cheap labor for growing industrial production.[20] Reformers increasingly promoted enclosure and associated "improvements" because of the new efficiencies they would deliver in a context of scarcity and a growing population. These new efficiencies were based on observable tendencies and trends, and they were not just concerned with increasing profit for landowners. They extended a new way of thinking about and governing the population, one that was based on the new fields of scientific knowledge analyzing all manner of social, economic, biological, and physical phenomena.

Writing about eighteenth-century liberal economics, E. P. Thompson acutely observed that "in the new economic theory questions as to the moral polity of marketing do not enter, unless as preamble and peroration" (1993, 202). The demoralizing and depoliticizing of decision making is the real power of liberal political economy, excluding as "sentimental," "naïve," or dangerous any counterclaim or voice of protest. Hardin criticizes this very same anti-economic sentiment: "every new enclosure of the commons involves the infringement of somebody's personal liberty. Infringements made in the distant past are accepted because no contemporary complains of a loss. It is the newly proposed infringements that we vigorously oppose; cries of 'rights' and 'freedoms' fill the air. But what does 'freedom' mean? When men mutually agreed to pass laws against robbing, mankind became more free, not less so" (1968, 6).

Despite intensive commercialization and exploitation, the Irish and European fisheries have avoided the implementation of private rights of access until now. It is not surprising therefore that these developments should have historical parallels with the enclosures of land in the late eighteenth century, when similar arguments about productivity and efficiency were being employed in a context of scarcity and, at that time, underproduction. For many independent fishermen who continue to be absorbed by the demanding activity of fishing, the costs in time, energy, and money to become "efficient" operators will be beyond them.[21] These individuals are the unfortunate but necessary collateral of the shift toward the "balanced," sustainable European fisheries of the future. The different and often mundane ways in which new regulations and requirements exclude fishermen have been usefully analyzed as "creeping enclosure" (Murray et al. 2010). This identifies enclosure as the less spectacular "process and function of multiple events and processes that need not be the result of a single regulatory moment." Ebby Sheehan, a fisherman and representative of the Irish Fishermen's Organisation, put this in his own words: "There are men in this industry now that are basically suicidal because of all the bullshit that we are made to adhere to. It's absolutely ridiculous, every day you come to the pier, there's either a fishery officer on board a boat or a marine surveyor officer about a boat. It's very hard for fishermen to keep focused on the job they are supposed to be focused on, which is catching fish" (Sheehan 2009, 3). As fishermen confront more pressures to improve the "efficiency" of their operations, the inevitable result will be the exclusion of those who are unable to adapt to the new requirements, adopt new fishing technologies and techniques, or enroll the support of scientists to monitor fish stocks and improve fishing practices.

This chapter has focused on the framing of European fisheries policy in terms of the "natural" limits of fish stocks and the need to "return" to a stable, ecological balance. Certain policy measures, such as the ban on discards and the liberalization of the quota system, emerge from this liberal formulation of the problem. The subsequent laissez-faire approach to governing imagines the market to be a self-regulating, natural sphere that policy makers should enable by removing any obstacles (such as fixed quotas) that impede its operations. This is important to keep in mind because the reform of the fisheries has also been characterized by a different formulation of the problem of overfishing and the proper role of the market in addressing it. In the next chapter, I will examine how ideas of the "green" economy translate into new ways of valuing and marketing the environmental performance of fishermen. Rather than representing a "return to liberalism," this specifically neoliberal rationality of governance marks a departure from liberal conceptions of natural limits and the laissez-faire policies that follow.

3

Stewards of the Sea

Neoliberalism and the Making of the Environmental Entrepreneur

The Green Opportunity

In May 2009 I attended a public meeting organized by the West Cork Development Partnership, a semi-state agency responsible for the promotion of development initiatives in the West Cork region of Ireland. The meeting launched a new round of submissions for LEADER, an EU-funded program run through the Department of the Environment, Community and Local Government.[1] The aim of LEADER is to improve the development potential of rural areas by providing financial support for local projects and businesses that are able to "achieve competitive advantage through the use of the area's unique image and identity."

Over 200 people filled a hotel conference room. Along with representatives of LEADER, several local councilors and politicians were on the panel of speakers. The first speaker from LEADER began by heralding the future as a time when we would all have "a chance to be masters of our own destinies, to control our own economic development." This optimism was based on the belief that the "factor-driven economy is past" and that the environmental agenda and information communications technology were providing new opportunities. "Clearly," he said, "the old ways aren't good enough. We need new ways of doing, new ways of producing."

Central to capitalizing on these new opportunities were the inherent values of a place like West Cork: natural beauty, an unspoiled environment, and a unique culture. In order to be eligible for funding from LEADER, potential projects had to demonstrate these qualities. Targeted areas for funding were diversification into non-agricultural activity, tourism, training, and information technology and skills acquisition. "Traditional" agricultural and fishing activities were not eligible for funding. Most of the evening was spent going through what appeared to be a long and complex application form. Finally, the audience was strongly encouraged to work with a development officer from LEADER to compose their applications.

After the event, I met James McCarthy, who was attending the meeting as a representative of the South West Fisheries Organization, the producers organization representing fishermen across the South-West of Ireland. He told me that fishermen were hopeful they could benefit from the scheme because it offered a way to "diversify their activities." Local fishermen were confronting reduced fishing opportunities and increased competition from global capture fisheries and aquaculture. While the Irish government's BIM has promoted a more market-orientated industry for at least thirty years, the structure of the fishing industry and the routes to market have meant that fishermen have largely focused on expanding their fishing capacity instead of "adding-value."[2] Until the reform of the CFP in 2002, government policies encouraged this trend by funding fleet expansion and better catch technology. Since then, the focus has begun to change as the reality of dwindling fish supplies, increased global competition, and European regulatory demands put more pressure on the industry.

The most recent report from BIM, "Capturing Ireland's Share of the Global Seafood Market," explains the need for the fishing industry to "diversify" their economic activity and generate value in new ways, through new economies of scale and new products.[3] It predicts that the Irish seafood sector will move from current annual sales of 700 million euros to 1 billion euros by 2020, and that it will increase full-time employment across the sector from 11,000 to 14,000. The reason for this optimistic outlook is the growing global demand for seafood. According to the UN Food and Agricultural Organization, demand for seafood is expected to grow by 42 million tonnes per year over the next decade (quoted in BIM 2013). While raw fish production in Ireland is going to expand with the development of the aquaculture sector, increasing volume production by 78 percent by 2020, the focus for the capture fisheries is the maximization of added-value through improvements in quality, marketing, and processing.

By emphasizing the global market as the "guiding light" of the industry, a different set of priorities opens up within the fisheries sector. This is reflected in the recent rebranding of the fishing industry as the seafood industry.[4] As Jason Whooley, CEO of BIM, stated in an interview,

> If we can begin to create a shift in focus from being a "stand alone" fisherman to seeing ourselves as a "supplier" selling his seafood to a major international market that has huge demand for this product, the opportunities for maximizing the return to fishermen become more tangible. *Fishermen have to begin to see their livelihood as a business with a major potential. The traditional routes for fishermen to bring their products to market have to be looked at in light of developing global trends if Ireland is to remain a strong player in the world seafood market.* BIM are keen to begin this debate and assist our industry to make the right decisions to ensure they don't see this major business potential lost. (Whooley 2011; emphasis added)

While part of BIM's strategy has targeted the development of the processing sector, another focus has been the introduction of quality and traceability labeling, including voluntary labeling and certification for environmentally sourced fish.[5] Projects like LEADER have supported the use of labels to differentiate food products on the basis of locality or quality, but BIM have focused on developing environmental accreditation for the fishing industry. The two-fold rationale for this is that fishermen who are fishing sustainably will be rewarded for their environmental practice, and the Irish fishing industry will be able to access high-end markets by differentiating their products from global fish products that do not meet the same standards. As the latest BIM report puts it, "Through our development initiatives, we will seek to grow the seafood sector by applying green economy principles that align the preferences of environmentally conscious consumers while maximizing renewable resources to reduce waste and input costs, and to embrace assured food production systems" (BIM 2013, 20).[6]

Although environmental accreditation is offered as an economic *opportunity* for individual fishermen, growing public and regulatory pressures on the fishing industry are also forcing fishermen to demonstrate their environmental credentials. A common remark during my research was that fishermen were losing the "moral battle."[7] As Sam Lee, the fisheries scientist working for the Marine Institute, told me,

> My own view is that fishermen have lost, or are losing this moral game of custodians of the marine environment. Joe public doesn't see that anymore. Twenty years ago sure, they were toilers of the deep. It was a respected business. Being a fisherman, there was kudos. In terms of people's views now, look, we read it all the time—and its complete bollox—but it's this populist view that they're towing metal beams across the sea beds and they're shooting seals and smashing coral reefs and they're catching dolphins. . . . In Dingle, or Rossaveal or Greencastle they still have that standing in the community, but in terms of the wider community I think that's gone, and I think that fishermen realize that. Whichever way you look at it, I think they always feel they're on the back foot, rightly or wrongly, from the greens, the oil prices, imports, you know. (Lee 2009)

In response to these pressures and the promise of added value, BIM is now providing fishermen with tools to "differentiate" themselves on the basis of their environmental performance. The Environmental Management System (EMS), launched in 2010 by BIM, is one such tool. It is a voluntary audit designed to help Irish fishermen "minimize their impact on the marine environment." Importantly, the EMS is also a way for fishermen to record and document their impact on the marine environment so they can demonstrate responsible fishing. The motivation for fishermen is that it will help them engage and negotiate with various state and nonstate actors, whether in terms of applying for international eco-accreditation for the fish they catch or in demonstrating they are in compliance with the regulations. Although BIM presents EMS as a transparent, “bottom-up” management tool designed to encourage and reward responsible fishing, the reality is that it involves ongoing work by fishermen with minimal prospect of any financial reward because of the opaque, technical, and costly nature of the global market and regulatory networks in which they seek recognition.

Recent literature has identified eco-labels and environmental accreditation standards as novel forms of neoliberal environmental governance that reflect the assumptions and goals of ecological modernization (Foley and Hébert 2013; Guthman 2007; Higgins et al. 2008; Klooster 2010; Ponte and Cheyns 2013). What is not so clear is why and how a policy tool like eco-labeling is different from other market-orientated measures, such as the individual transferable quotas outlined in the previous chapter. These differences are important because they can help us differentiate liberal and neoliberal responses to ecological crises and the different subjectivities and “natures” that are being constituted in the process. In this chapter, I consider how neoliberal policy makers have moved beyond “natural” biophysical and economic limits by actively constructing the market as a site of competition, the individual subject as an environmental entrepreneur, and nature as informational content. These three elements cohere around the evaluation and commodification of “environmental performance” and the supposed liberation of economic growth from material concerns; fishermen enter into competition not over their capacity to extract fish but over their environmental “performance,” which is measured within highly uneven transnational accreditation networks and markets.

The previous chapter documented how liberal political economy operates within the presumed limits of bioeconomic nature. This chapter examines the distinct and novel character of *neoliberal* political economy—the attempt to overcome limits through the ongoing cultivation of innovation and competition. It first outlines how the EMS attempts to provide a more empowering form of environmental regulation that bypasses the inflexible, bureaucratic apparatus of the state by directly harnessing individual innovation and enterprise. This approach resembles a broader tendency in areas of work, healthcare, and education, for example, where individual performance has become subject to new forms of measurement and control. I draw on Foucault's distinction between liberalism and neoliberalism to explain how this technology diverges from the policy measures examined in the last chapter. Although both liberalism and neoliberalism share the belief that the market is the site of verification for governing society, Foucault pinpoints a transformation in liberal thought in the early part of the twentieth century. Central to this transformation was a move away from laissez-faire policies to a more interventionist strain of liberal thought that emphasized the logic of competition as the means for achieving all manner of social and environmental goals. The harnessing of competition through new technologies of measurement and assessment thus becomes a driving motivation of neoliberal policies and the basis for a new apparatus of governmentality.

The scope of my analysis then expands to consider how the EMS fits within new transnational "sustainability networks" that illustrate the changing roles and relationships between the state, market, a range of nonstate mediators, and individual fishermen (Ponte and Cheyns 2013). These governance networks are supposed to provide a more participative, transparent, and less prescriptive means of achieving environmental goals (Schmitter 2002). But this approach presupposes that all "stakeholders" are already equal. Behind the optimism and positive rhetoric surrounding these new modes of governance, what we find is the emergence of new ways of identifying, measuring, and commodifying nature that redraw the boundaries of participation in the green economy.

The New Ethos of Environmental Governance

Physically, the EMS consists of a booklet, a little bigger than A4 paper, laid out in fourteen chapters or sections covering every aspect of a fisherman's environmental, economic, and community performance, including fueling operations and maintenance, hygiene management, fishing practices, food safety and quality, waste management, occupational health and safety, communication and public perception, and sustainable fisheries management and research. While the ultimate aim of the EMS may be accreditation, fishermen are also encouraged to become more fuel efficient (saving money and reducing their impact on global warming) and to improve the quality of their catch through better handling on board. The manual contains suggested actions, timelines, worksheets, and space for notes. The layout of the manual is always the same, but each individual fisherman decides independently what aspects of his activity he wants to assess and how. There are no set requirements in the EMS; the objective is simply to allow individual fishermen to record what they do. One of the assumed advantages of the EMS is that it operates at the level of the individual, the "smallest unit of management." Jean O'Sullivan, who spearheads the project for BIM, enthusiastically endorsed the system when I spoke with her. She cited the success of an Australian seafood EMS that "will be in all third level education books as a model for environmental management."[8] The particular project concerned the work of an oyster farmer in a small estuary in Swanport, South Australia. The project, which won a United Nations Seafood Environmental Award in 2005, was considered a success because the oyster farmer had extended the scope of his EMS *beyond* any objectives laid out in the manual. He even managed to change how other farmers managed run off into the estuary, thereby affecting the whole ecology of the area. The oyster farmer received recognition in a "type of OBE" (Order of the British Empire) for services to the environment. O'Sullivan emphasized the value of a tool like the EMS for providing the space for individuals like this "to take the initiative." This contrasts with previous models of environmental management that tend to prescribe what should be done within rigid regulatory frameworks. She told me,

> The only way you'll drive change is by empowering the institution to change. If it's top down, it's just another layer of legislation: you must do this because the law says so. It always drives the lowest common denominator, whereas what you really want to do is to try and increase the environmental performance, not just for the environment's sake but for the fisherman's sake. So their outline is that you actually put the fisherman central to the process. It's not about the environment,

> which can sound very callous, but if you put them there first and say *a function of their livelihood is protecting the environment* that's a much better reason to respect the environment or maximize the environment than saying it comes first and we don't care about your livelihood. So call it . . . looking at the human element of what they do and how they do it. (O'Sullivan 2009; emphasis added)

O'Sullivan cited another example of a fisherman in the south of Ireland who had recently decided to "lead" on the issue of accreditation: "he's suddenly decided, you know what, through this process I want to drive x, y, and z rather than sitting out at the committee level and it being the lowest common denominator of the committee. . . . He's decided, you know what, this has given him the tools, or abilities, or method of going you know what, fuck them all now I'm going to lead them on this." Simon Casey, a colleague of O'Sullivan's, went on to discuss how this demonstration of leadership could work as a catalyst for the rest of the industry: "if he finds better grazing, the rest of the flock will follow. So that's why you need champions" (Casey 2009).

A second assumed advantage of the EMS is its flexibility: it enables fishermen to gradually build up a record of their activity and impact on the marine environment. This record effectively equips them with a language to represent themselves when the time comes. The emphasis here is on empowering individual fishermen to take control of their fishing practice to better situate themselves within increasingly demanding and uncertain regulatory and market contexts.

> It doesn't matter if they're in Sellafield or out on their boat, the layout is universal, so that it's easier to communicate. Now there are a couple of things with the EMS. Okay, they're managing their environmental performance and things come and go in roundabouts a lot. First of all they need something to improve their compliance, or to improve their market advantage. But its funny, its not just one thing, and things will change by the season if you know what I mean, even the day, it depends on what they're dealing with. *But by going through this they're beginning to acquire a language.* (O'Sullivan 2009; emphasis added)

The "beauty" of the EMS, O'Sullivan told me, is that it can be adapted to the individual operator but at the same time provide a "universal" basis for accreditation. It manages to be both uniform and unique.

The EMS bridges the apparent gap between ensuring compliance and consensus and enabling fishermen to be creative and adaptive within their own locality and environment: "No two EMS are the same," O'Sullivan confirmed. Unlike the overly prescriptive, top-down regulations of the past, the EMS allows fishermen to acquire their own "voice":

> Their (fishermen) objective is to generate a positive and public image for our business and the industry. *How they do it should be their business provided it is done within the controlling legislation framework and the norms of public acceptance.* But you give them the control rather than making it prescriptive: you must do this and you must do that across the fourteen aspects. *You give them the power to be creative.* (O'Sullivan 2009; emphasis added)

O'Sullivan told me about a meeting she had with a fisheries cooperative in the West of Ireland. They had wanted to work on a marketing plan for their crab by using the EMS, which they had heard of through a third party. She asked them what made their crab different. They did not know what to say; they had no way of showing or proving how they caught the crab, how pure the waters were, or how healthy the fish stock was. By engaging with the EMS, committing to better fishing and handling practices, and, most importantly, documenting these practices, the fishermen could provide tangible proof of their environmental fishing practices and unique product. They could have a "book of good deeds," as O'Sullivan put it. It was not just a matter of translating this documentation into a label, though this was important, but of raising the confidence of the fishermen by "giving them something to bargain with." At the end of the meeting, which had begun as a means to gain added value for their crab, the fishermen raised concerns about a new European environmental directive that could affect their ability to access the fisheries.[9] By having the documentation provided through the EMS, however, they would be able to engage in future consultations and hopefully defend their position as "environmental fishermen." She told me the EMS was necessary "so they [fishermen] can go to local meetings and so on, and say we're the fishermen and we're not as bad as people are making us out to be and here's the proof." She continued,

> A big factor is that . . . it has got to a point where in mixed company they are embarrassed to say they are fishermen because of the negative press. They are saying . . . the criminal bill is there . . . so basically they feel really harassed in the media and they feel if this (EMS) is something I can do to change my image, and it's not like superficial, it is about actually saying I am a primary producer, I produce food . . . you know? Yes, I work in a very regulated environment and there is a criminal bill there, but that doesn't mean I'm a criminal. (O'Sullivan 2009)

This is why O'Sullivan considered the EMS to be much more than a marketing tool or a top-down policy innovation being imposed on fishermen. For her, the EMS would "empower fishermen to be confident and articulate," to defend and differentiate themselves as sustainable and productive stewards of the marine environment. "It allows them," she said, "to stand up and say this is who I am, this is what I'm doing, this is where I'm going." She described it as an "ethos" and a "philosophy."

Jean also admitted that the EMS was "time rich." This was not just because of the extra activities that an individual might have to take up and monitor (such as waste management, recycling old fishing gear, or engaging with scientific data collection) but also because of a more general need to be open and responsive to a whole range of dynamic social, environmental, and economic factors. She suggested, for example, that fishermen needed to realize that issues such as offshore renewables were their responsibility.[10] In this sense, the EMS was not limited to any specific criteria, a point she kept returning to. She distinguished the EMS from other standards that set specific criteria and were thus "static." Indeed, she admitted that it was not a standard at all because it did not promise anything in terms of biological sustainability or better access to markets. The EMS was a tool that inserted the individual fisherman into an open-ended and constant process of improvement and engagement.

At the same time, the EMS emerges in response to a particular context. As her colleague Simon Casey made clear, it is no longer enough for fishermen to just mind their own business: "essentially the international bar is rising and all fishermen have to rise with it and those who don't are going to miss out and be left behind" (Casey 2009). Although the EMS itself may not be prescriptive, it has emerged in response to an economic and regulatory context that is prescribing new limits and demands on fishermen. In this context the rhetoric of "empowerment" and "freedom" is juxtaposed with the "limitations" and "demands" of the market and environmental regulation that are forcing fishermen to "adapt." The challenges of competing in the global market were summarized by John Nolan, the head of the Castletownbere Fishermen's Co-op:

> One of the products we're coming up with at the moment is panga fillets from Vietnam. They're two euro a kilo for the fillet delivered to any mainland European city. They are produced from freshwater farming in Vietnam. . . . You also have Nile perch fillets coming in from Uganda, Lake Victoria. You have Tilapia fillets coming in from Indonesia, never mind the hake coming in from South Africa and the hake from Chile. We sold a boat recently to a company in South Africa and the hake quota for the whole of the Irish fleet is about 1200 tons, and one boat in South Africa has 30,000 tons of a quota. . . . So Europe is swamped with imported fish and we are just left to survive and we're not surviving. The industry is in terminal decline. Sixty percent of the fleet are in interest only with the bank. Last year we had a huge oil crisis. In the last five years, forty boats took decommissioning [and] sold to get out of the business. So we would be employing four hundred full-time fishermen at sea five years ago. We now only employ 180. (Nolan 2009)

In the face of such market pressures, the EMS is literally presented as a means for fishermen to defend themselves against global competitors. As Casey put it,

> *It (EMS) is a way of arming themselves against an invasion like that of cheap products from countries with cheap labor.* A lot of those products are probably to do with aquaculture, like catfish, tilapia. Then you've got imports of cod from Iceland, [the] Faroes, and flights going into Cork. Part of the argument is 'Ban the imports, where is the EU doing anything about this?' *But there's another side of the coin: it's a free market economy so you can't stop that happening. What you have to do is fight against it.* And this is one way of fighting against it, as is a standard, as are the NGOs convincing consumers to demand a particular level of responsibility for a product. (Casey 2009; emphasis added)

The globalized market is the inevitable backdrop to the emergence of the EMS as a "springboard" to accreditation and traceability standards. Rather than opposing or limiting global market competition, the EMS and corresponding global accreditation schemes represent a clear and explicit attempt to harness and expand the logic of competition to achieve both economic and environmental objectives. By pitting fishermen against their "less sustainable" competitors at home and abroad, the EMS attempts to resolve economic and environmental problems besetting the fisheries sector by extending and transforming the market rather than regulating it. In response to a question I raised about cheap, imported fish products being a principal obstacle to a sustainable fishing industry—a question that animated many of the fishermen I had spoken to—O'Sullivan replied:

> *Fishermen as a livelihood, other consumers, they are accepting globalization of products.* The tilapia, warm water prawns etc. It's accepted as the globalization of commodities. But that doesn't help the small fishermen, *so basically it's only in their power to stop it. Now it's up to us to try to enable them to provide a standard that they can say this is differentiated, as in it's done by quality, provenance, responsible practice, or by "I have a seafood EMS."* (O'Sullivan 2009; emphasis added)

Analysis of eco-accreditation and eco-labeling has rightly pointed to the market logic that underpins such policies: the enclosure and commodification of new forms of activity and knowledge, and the enrolling of individuals as the instruments of environmental "improvement" (Guthman 2007).

What appears to be missing in these accounts of contemporary neoliberal environmentalism, however, is why and how this mode of governance might differ from liberal conceptualizations of the market and the political economy that informs it. Much of the literature on the contemporary neoliberalization of nature tends to avoid making any distinctions between the neoliberalism of today and the liberalism of the past; the "neo" in neoliberalism is assumed to indicate the return of liberalism rather than any transformation in liberal thought and governmental practice (Amable 2011; Lemke 2002). One of Foucault's great contributions to our understanding of liberalism is his method of grounding liberalism within the specific historical and sociomaterial contexts it has taken shape. By providing a historically rooted account of the development of liberal thought from the end of the eighteenth century to the 1980s, he illustrates important distinctions between liberal and neoliberal modes of government.[11] Although both sought to discover and implement an economic rationality for managing society, the way they conceived of the market and its function in governing social life were quite different. It is worth looking briefly at how this reconceptualization of the market and social regulation differs from the classical liberal account and how this distinction informs the post-Malthusian optimism of ecological modernization.

In Foucault's analysis, a new strain of liberal thought emerged in the early to mid-twentieth century in response to what was perceived as an inherent flaw in classical liberal economic thought: the "naive" formulation of the market as a "natural," self-regulating sphere that operated outside the authority of the state, a belief that produced the laissez-faire approach to governing examined in the previous chapter. In addition to "mystifying" the economy as a domain of activity outside other areas of social life, the market's identification with freedom of exchange meant that its main function was to ensure the balance of supply and demand. In contrast, *neo*liberal thought reconceptualized the market not as a site of exchange but as a *site of competition*: it is competition between individuals that provides the necessary and continuous friction for mobilizing society toward certain goals (Amable 2011; Dardot and Laval 2013). What characterizes neoliberal thought, then, is the denaturalization of the so-called "free" market that operates at the limit of the state and the cultivation of competitive market relations that can, and must, enter into all forms of social activity, not just economic production.[12]

This reconceptualization of the market in terms of competition necessarily requires the state to take on a new role. Laissez-faire policies were understood in terms of *allowing* the market to function freely by facilitating the circulation and exchange of commodities and labor. The neoliberal state, in contrast, must take more deliberate and active steps to *construct* the market so that competition can thrive in areas where it may not have existed previously (Burchell 1996). Once the market becomes synonymous with the logic of competition, the role of government is to ensure that competition always takes place "in such a way that the market is always maintained and that the principle of equal inequality produces its effect" (Donzelot 2008, 124). "Equal inequality" refers here to the need for government to ensure a constant level of *relative* inequality between individuals. Rather than creating liberalism's level playing field, a regulatory framework that treats everyone the same, the role of neoliberal state intervention is to cultivate artificial forms of inequality in order to encourage and incite particular forms of behavior: "Competition has an internal logic; it has its own structure. Its effects are only produced if this logic is respected. It is, as it were, a formal game between inequalities; it is not a natural game between individuals and behaviors" (Foucault 2008, 120). At the heart of this rationality is the assumption that as long as an activity can be measured and thus compared, all areas of social life can be subject to the same productive logic of competition. Whereas liberals defend the principle of noninterference or laissez faire, neoliberals are preoccupied with "permanent vigilance, activity and intervention" (Foucault 2008, 132). Neoliberal governmentality therefore opens up a more extensive and intensive role for the state through the multiplication of individual performance indicators and measures of ability: "It involves generalizing [the economic form of the market] throughout the social body and including the whole of the social system not usually conducted through or sanctioned by monetary exchanges" (Terranova 2008, 243). The best remedy for economic (and ecological) crises is therefore not ineffectual efforts to balance supply and demand but the active harnessing of the competitive spirit to bring about desired transformations in society (Lazzarato 2009). As Dardot and Laval write,

> The market ceases to be the natural "air" wherein commodities circulate unhindered. It is not an "environment" given once and for all, governed by natural laws, ruled by a mysterious principle of equilibrium. It is a regulated process employing psychological springs and specific skills. *It is a process that is not so much self-regulating (i.e. leading to perfect equilibrium) as self-creative, capable of generating itself over time.* (Dardot and Laval 2013, 123)

While liberal political economy assumes and operates on the basis of "natural" limits—both bio-physical and human—neoliberal political economy attempts to shake off such restraints by emphasizing the unrealized potential that exists within both human and nonhuman nature, a potential that can be realized through the permanent cultivation of competition.

In her book *Life as Surplus*, social theorist Melinda Cooper identifies this neoliberal turn in the environmental arguments and debates of the 1960s and 1970s. She describes how the return of Malthusian arguments and gloomy prognoses precipitated a novel response that resonated with postwar neoliberal ideas seeking to move beyond the limitations of classical liberal thought. While acknowledging the environmental limits to industrial production, these new visions of postindustrial growth emphasized the immaterial, technological, and innovation-based aspects of economic activity that promised not only humanity's release from the "natural" limits of the planet but also a solution to looming environmental crises. Cooper goes on to examine the specific case of the biotech industry, but her analysis can be more generally applied to the entwined projects of ecological modernization, sustainable development, and the "green" economy that are only now being analyzed in these terms.[13]

While these projects cover a broad set of policies and approaches, they share an understanding that environmental problems (such as overfishing) do not contradict the possibility for continued economic growth. Rather, efforts to manage environmental problems can provide the basis for new forms of "green" growth through the construction of new green values (Beck 1992; Brockington and Duffy 2010; Brundtland et al. 1987; Hajer 1995, 1996; Mol and Spaargen 2000; Murphy 2000; Sullivan 2013; Warner 2010). Ecological modernization thus signals a departure from the Malthusian preoccupation with *balancing* human activity with limited biophysical resources. Instead, new innovations (such as eco-labeling) can be employed to transform the economic base from one that is reliant on exploiting limited, natural resources, to one that thrives on the "immaterial" inputs associated with "added-value", namely informational or symbolic content (Organisation for Economic Co-operation and Development 2011). Eco-labels and accreditation schemes offer a good example of how this rationale takes shape through new technologies and cultures of valuation.

If individual transferable quotas allow the free circulation of quotas among fishermen, incentivizing and rewarding those who are more efficient at catching fish, auditing technologies like the EMS and associated eco-labels encourage competition between fishermen in terms of their environmental *performance*. The EMS is a specifically neoliberal technology because it measures and evaluates individual performance in order to compare it with others. This competition is understood to incite desirable types of behavior (environmental "stewardship") and also open up new areas of economic value that do not rely on intensified levels of resource extraction. Significantly, the environmental performance of an individual fisherman is not limited—there is always room for improvement and the auditing process does not stop. While the ban on discards and the introduction of ITQs encourage fishermen to adopt efficiencies that are both prescribed by and assessed in relation to the biological health of the fish stocks, the EMS and accreditation schemes are not (in theory) limited by anything except the willingness of the individual fisherman to demonstrate his commitment to his environment. It is therefore important to recognize that these transformations in neoliberal thought represent powerful new currents of environmental governmentality that become crystallized in the figures of the neoliberal subject and neoliberal nature. In liberal environmentalism, *homo economicus* works within the limits of biophysical nature, making optimal decisions with scarce resources. In neoliberal environmentalism, by contrast, the environmental *entrepreneur* must not only respond to, but also create, new opportunities for himself within a malleable and dynamic ecology and market.[14] As sociologist Jacques Donzelot concludes,

> If what counts is no longer, in the first place, the man of exchange, the man of need and consumption, but the man of competition, the man of enterprise and production, then we should encourage everything in him that partakes of this spirit of enterprise and place our reliance in man as entrepreneur: as the entrepreneur of an economic activity, of course, but also as an entrepreneur of himself—the wage-earner is only ever someone who exploits his own human capital—and as a member of a local collectivity taking care for the maintenance and increase in value of their goods. (Donzelot 2008, 124)

A considerable body of literature documents the rise of both an "enterprise culture" and a "culture of responsibility" across different sectors of civil society (Burchell 1996; Read 2009; Rabinow and Rose 2006; Rose 2006; Rose and Miller 1995). In the fields of work, healthcare, and education, individuals are increasingly "empowered" to take on responsibility for their future, "to be understood and targeted as an active participant in the activity of work, not merely as an instrument of production but as a person: a human being realizing his or her self through work" (Rose and Miller 1992, 430). In order to harness the dynamic power of competition to achieve social or environmental objectives, individuals have to take on more responsibility for their activities and be encouraged to be autonomous and independent.[15]

At the same time, competition requires the creation and insertion of new forms of measurement and assessment in order to differentiate individuals: although individuals may be incited to compete against one another (in the search for "champions"), they do not compete within a Hobbesian state of nature. "Contingency means lack of limits rather than lack of order," writes sociologist Luigi Pellizzoni. "Better: disorder, as a positive, enabling systems condition, can be handled by carving out provisional room for purposeful manoeuvre" (2011, 797). In other words, there are rules to the neoliberal game that are carefully designed using the same calculative logic that undergirds liberal political economy. Instead of shaping individual behavior through clearly delineated material incentives and disincentives, however, neoliberal governmentality seeks to enable individuals to seize opportunities when they appear. As Michel Callon writes, the new entrepreneur is "defined . . . by the ability to set his own goals, to determine and calculate his interests, to stand back from the world surrounding him, to instrumentalize it, to be utilitarian, selfish and opportunistic" (2007, 142). In the case of the EMS, this activity includes constant monitoring of everyday activity and the translation of performance into different spheres of negotiation, including international accreditation.

When we spoke, O'Sullivan emphasized that the EMS was "all about the journey, not the destination." Just like the oyster farmer in Australia, the environmental entrepreneur always has more to offer: more areas to improve in, more people to engage with, more markets in which to sell his products. Neoliberalism creates "a new form of 'responsibilization' corresponding to the new forms in which the governed are encouraged, freely and rationally, to conduct themselves" (Burchell 1996, 29). The EMS and associated accreditation schemes are a good example of how this neoliberal logic of "responsibilization" and entrepreneurialism has extended into environmental management. It also provides us with important insights into how the neoliberal response shapes new roles for the state. Casey described this transformation in our conversation:

> This is a process driven by the individual, whether going for the MSC [Marine Stewardship Council] standard or the BIM standard or any

other third-party standard, or for whatever purpose. We want them to come to us and say what can you help us out with because of x, y, and z. It might be market, green issues, better community presence, or appearance. We want them to come to us and then get the mentors in to instruct them how best to do it, and to talk to their peers and then provide what advice we can, so rather than top down we're trying *to assist from the side* as it were and bring them on so they're growing from the bottom up. (Casey 2009; emphasis added)

Casey had no doubts that competition from overseas for cheap fish products was not only undermining the economic viability of Irish fishermen but also forcing them to exhaust what stocks remained. His answer, though, was not to abandon the idea of competition but to harness it in another direction. The liberalization of trade has meant fishermen around the world are competing on a level playing field. The introduction of eco-labels punctuates this level playing field through the artificial differentiation of products: the creation of secondary markets in "environmental" seafood. The EMS aims to insert fishermen within "a set of material and technical devices, incentives and forms of organisation that have no logic other than creating ecological niches in which 'economizing' human agents can survive and even prosper" (Callon 2007, 142). This is not about "allowing" things to happen "naturally" but recognizing the need to intervene and artificially create the conditions in which competition can take place.

The EMS is intended to "empower" individuals to engage more directly with international NGOs and market actors to resolve shared problems and accomplish shared goals using a shared language. As O'Sullivan was keen to stress, this model is intended to allow fishermen to bypass the bureaucratic and hierarchical structures of the state and gain direct recognition and reward for their environmental performance. The ambiguous role of the state in this process is encapsulated in Casey's idea of "assisting from the side." The state withdraws from direct regulation of fishermen but applies greater pressure on their everyday lives through regulatory environmental norms and individualized auditing technologies.[16]

This tension between deregulation and intervention is once again founded on the supposed freedom of individual fishermen who are being encouraged to adapt to changing regulatory and economic demands. Illustrating this tension was an exchange that took place between O'Sullivan and Casey during our meeting:

> JS: [They want to] generate a positive and public image for their business and the industry. How they do it should be their business, provided it's done within the controlling legislation framework and the norms of public acceptance. But you give them the control rather than making it prescriptive, you must do this and you must do that across the fourteen aspects. You give them the power to be creative.
>
> SC: But if they were using it for a particular standard, like MSC, there is an element of prescriptiveness that creeps into the design of the EMS—to serve [as] the background for, say, the MSC.
>
> JS: This is a voluntary project—if they want to they can.
>
> SC: If they want the gold standard, they have to be prescriptive in certain areas because otherwise the standard isn't a standard.
>
> JS: Well, well, I would say you make the outcomes clear, you give them targets, . . . [but] you shouldn't make it overly prescriptive. I don't know how many operators we have from the pelagic right down to the inshore. Being prescriptive won't suit them all. But if you gave them an outcome or a target it is more friendly.
>
> SC: Well, your EMS is designed to be nonprescriptive, but the standard, any standard is designed to be prescriptive, so the two of them have to marry up.

In addition to shifting responsibility for the fisheries "down" to the individual level, technologies like the EMS also fit within emerging supralocal and transnational scales of governance. The EMS presents individual fishermen with a tool or "prosthetic" that is supposed to allow them to connect directly and transparently with global accreditation schemes and consumers of sustainably caught fish. Rather than fulfilling this promise, however, the new forms of transnational governance connected to the EMS are both opaque and unaccountable. As Dardot and Laval write, "Evaluation has become the key to the new organization, and crystallizes all kinds of tensions, if only that which results from the contradiction between the injunctions to creativity and risk-taking and the corporate appraisal that descends as a reminder of the real balance of power in the enterprise" (2013, 199). The Marine Stewardship Council (MSC) accreditation scheme and its cooperation (or not) with the EMS illustrate this new balance of power.

The Marine Stewardship Council: Transnational Governance and the Institutional Void

Jean O'Sullivan said that one of the main objectives of the EMS is to enable fishermen to achieve the "gold standard" of environmental accreditation in the fisheries: the MSC accreditation. The multinational company Unilever and the international NGO World Wildlife Fund established MSC in 1997. As one of the first worldwide certifying bodies, the council was established to raise consumer concerns about declining fish stocks. In 2010 the MSC gained media attention through its association with the documentary *The End of the Line*, based on the book by Charles Glover. The film was part of a wider campaign to raise awareness about the problem of overfishing and encourage people to buy environmentally sustainable fish. The final scene in the film shows Glover calling a well-known restaurant in London demanding to know where their fish comes from. Around the same time, "Fish Fight," the high-profile campaign to end discarding in European fisheries, was also taking place. Led by the well-known television chef Hugh Fearnley-Whittingstall, "Fish Fight" also sought to raise consumer awareness by encouraging individuals to ask their supermarkets how they sourced their fish. In the wake of the campaign, all major supermarket chains in the UK changed their sourcing policies on fish, with Marks and Spencer, and Walmart in the United States, going a step further by committing to stock only MSC-certified fish. Despite the existence of other seafood eco-labels, the MSC remains the dominant player in the field, giving it a near monopoly in the "sustainable fish" market (Gulbrandsen 2009; Foley and Hébert 2013). To fishermen and the organizations that sell their fish, MSC certification is seen as a golden ticket to the major global retailers. The problem for most fishermen and fisheries is that achieving MSC accreditation is difficult because of the costs involved, the level of information and data required about the fishery, and the need for strong management structures.

The MSC assesses the sustainability of individual fisheries on the basis of thirty-one performance indicators principally related to the biological sustainability of the particular fish stocks being accredited. This means that individual fishermen cannot apply for accreditation on their own: all the vessels involved in a particular fishery must be committed to the accreditation process, otherwise there is no way of assuring that the stock is being fished sustainably. The process can begin only when all the fishermen targeting a specific fishery (a species of fish that is associated with a particular marine area) come together and decide to enter into assessment. This coordination presumes a considerable degree of biological, economic, and social coherence that does not pertain to most sectors of the Irish fisheries. Assuming that a fishery has reached this point, the MSC guide estimates it will take another seventeen months to achieve accreditation.

Once a group of fishermen have identified themselves and the fishery they wish to gain accreditation for, a "fishery client" must make a formal application to the MSC for assessment. The "fishery client" can be a government agency or fishing industry association, but it must have some sort of legal standing so it can sign contracts on behalf of its members. In Ireland the "fishery client" would generally be BIM, a state agency. The fishery client is responsible for identifying the "unit of certification" (what vessels and species are being assessed), paying the cost of the accreditation process, and the implementation of any conditions after the accreditation has been achieved.

The fishery client is also expected to identify an independent "third-party" certifier accredited by the MSC. These independent certifiers are themselves certified by Accreditation Services International (ASI), the sole provider of accreditation services to the Forest Stewardship Council, the Marine Stewardship Council, and the Aquaculture Stewardship Council. The authorized independent certifier is paid a fee by the applicant fishery in order to carry out the assessment and issue certification. This fee is worked out between the fishery client and the certifier. The independent certifiers are thus private commercial enterprises, and applicants are advised by the MSC to seek tenders from different certifiers in order to get the best deal.[17] The cost of the preassessment, full assessment, and then annual audits is estimated to be between 15,000 and 120,000 dollars.

In order to speed up the assessment process, applicants are encouraged to have as much biological, economic, and regulatory information as they can for the assessment team. Often this information is not readily available because of the limited resources available for carrying out such research. As mentioned in the previous chapter, the growing need for scientific data in the fishing industry is increasing demand for scientific consultants. Applicants are also required to engage with various other "stakeholders" who will be interviewed as part of the assessment process. These stakeholders include government agencies, environmental groups, other fishermen, community organizations, the commercial/postharvest sector, and scientists.

The three main principles that the MSC requires of a fishery are: sustainable fish stocks, defined as a fishery that can operate *indefinitely*; minimizing environmental impacts to ensure that the ecosystem on which the fish stocks depend is maintained and preserved; and compliance with all local, national, and international laws, including a management structure that is capable of responding to changing circumstances. The MSC recognizes that each fishery is different and warrants a unique assessment and certification process. The fishery does not receive any explicit guidance or support about how to gain certification by the MSC or the certifier: "the certifier's role is to offer guidance and make clear to you the required outcome rather than prescribe actions that should be taken. *The decision is therefore yours to make on how to achieve the desired outcomes*" (MSC 2011, 20; emphasis added). However, the fishery must gain a score of 60 or more in each of the 31 performance areas and a score of 80 or more in the 3 principal assessments to secure certification.

If the certificate is awarded, the fishery must then take steps to ensure that the MSC label is used on its catch and that they benefit from it. Implementation involves engaging with the certifier and the postharvest sector all the way down to the final retailer to ensure there is a transparent "chain of custody" in place. Onshore processors, distributors, and retailers also must comply with an eco-label licensing procedure. The MSC certification guide indicates that both these processes carry "budgetary implications" (MSC 2011, 26). The fishery is accredited for five years, provided that it complies with the conditions set down by the certifier and that it passes surveillance audits carried out by the certifier. These audits can be carried out at any time and must be funded by the fishery or fishery client. The funds to cover these audits must already be in place for the duration of the five years.

The cost and time involved in achieving MSC accreditation depend on the nature and complexity of the fishery. As the MSC emphasizes, "better organized" fisheries go through the process more smoothly and quickly. The problem for many fisheries in Ireland is that they are not well organized: they lack readily available biological information on fish stocks, clearly defined participants within specific fisheries, or the ability to control and respond to changes in physical, legislative, or economic environments. Nor, in general, do they have access to financial and administrative resources.

It is therefore no surprise that while the MSC has awarded four thousand certifications to fisheries around the world, it has only issued three in Ireland. All three of these have gone to the highly capitalized pelagic fishing fleets. These small fleets consist of large, modernized vessels exploiting fish stocks that are relatively sustainable owing to their high volume and coherency in terms of swimming in large, unmixed shoals.[18] It is more complex, and thus costly, for example, for the demersal fisheries to undergo assessment because they generally target mixed fisheries. Besides the biological and socioeconomic composition of the fishery, the demersal fisheries also face the problem of regulation because most of the fishing takes place in European waters in which the Irish government does not have sole jurisdiction: for a fishery to be certified by MSC, there must be an authority capable of regulating the activities of boats within the particular fishery. It is therefore almost impossible for any group of Irish fishermen targeting whitefish to comply with the three principles required for the MSC certification. The limitations of the EMS and the MSC offer a good example of the gap between what exists and what is required in the emerging regime of transnational "differentiation."

The assumption behind the EMS and associated MSC accreditation is that this process is more transparent (involving direct negotiation) and democratic (involving direct participation in order to address particular issues) than traditional forms of state regulation. The problem is that the EMS is just one element in a transnational "sustainability network" (Ponte and Cheyns 2013) that helps "comprise a re-scaling of environmental governance in which multiple stakeholders in commodity networks, including social movement actors, private sector producers and processors, retailers, and consumers interact in the governance of a commodity network" (Swyngedouw 1999: 1999). These emergent networks and alliances have been described as new forms of governance-beyond-the-state:

> Governance-beyond-the-state refers in this context to the emergence, proliferation and active encouragement (by the state and international bodies like the European Union or the World Bank) of institutional arrangements of "governing" which give a much greater role in policy-making, administration and implementation to private economic actors on the one hand and to parts of civil society on the other in self-managing what until recently was provided or organised by the national or local state. (Swyngedouw 1999: 1992)

Voluntary certification systems such as the MSC offer a good example of how transnational networks of government and nongovernmental agencies and institutions, including environmental groups, third-party accrediting bodies, state agencies, retailers, and, ultimately, consumers, are transforming how environmental problems are being governed. There is a clear shift in regulatory/institutional decision making away from the national scale, both upward to the supranational and downward to the scale of the individual. This form of governance demonstrates a decline in formal representative politics and the supposed rise of a highly developed, well-informed civil society operating through the free and open negotiation of different interests (Blowers 2003; Drummond and Marsden 1995; Kenis and Lievens 2014; Kooiman et al. 2008; Klooster 2010). However, "the status, inclusion or exclusion, legitimacy, system of representation, scale of operation and internal or external accountability of such groups or individuals often take place in non-transparent, *ad hoc*, and context-dependent ways" (Swyngedouw 1999, 1999).[19] The reality is that this governing process is more likely to be characterized by opaqueness and hierarchy than transparency and equality (Eden 2009).[20]

As O'Sullivan reiterated about the EMS, fishermen are free to decide what part of their environmental performance they want to manage without the normal, top-down prescriptions of state regulatory authorities. This apparent lack of prescription carries through to the MSC, where the virtue of the accreditation process is that each fishery is singular and specific and therefore must engage in an ongoing process of negotiation with the scientific community and other "stakeholders." In reality, the lack of guidance or clarity regarding requirements makes the process more opaque and uncertain.[21] Just five years ago, for example, the MSC was presented as the best indicator of sustainable fish stocks, but recently it has come under criticism from a group of leading fisheries scientists and international NGOs (Ainley et al. 2010). In an article in *Nature*, a team of six scientists supported by the Pew Foundation and Greenpeace argued that the MSC needed more stringent standards and stricter interpretation of its rules. The authors also pointed out that between 2000 and 2004, the MSC "boomed" as it became a commercial interest, culminating in Walmart's decision to sell only MSC certified fish. The article stated that the MSC had to alter its process of assessment in order to avoid a potential financial incentive to certify large fisheries. It pointed to the low accreditation of low-impact, inshore fisheries. The same year, another controversy emerged as the Pew Environment Group, an influential international NGO, criticized the MSC decision to award accreditation to the Antarctic krill (MSC 2010).

Sociologist Maarten Hajer has described the ambiguous and opaque nature of such contemporary forms of governance as the "institutional void," a situation in which "there are no clear rules and norms according to which politics is to be conducted and policy measures are to be agreed upon. To be more precise, there are *no generally accepted* rules and norms according to which policy making and politics are to be conducted" (2003, 175). At the heart of the "institutional void" is the supposed existence of a neutral space that is not monopolized by any one particular actor; all are equal partners working toward mutually agreed ends (de Angelis 2007). The institutional void opened up by these new forms of transnational governance does not remain open to all, however. It is a site dense with power relations: although there is an implicit assumption that eco-labels can function as a transparent tool mediating between "responsible" fishermen and "responsible" consumers, it is clear that there are many actors mediating this relationship and they are not all equal.[22] Jean O'Sullivan's description of the crab fishermen in Mayo is indicative. Under pressure from environmental directives from the European Union, they now must "demonstrate" their relative impact on the marine environment according to criteria set by external, transnational scientific institutions. Unelected, nonstate actors assume a greater role. Unilever, the company that established the MSC accreditation in collaboration with WWF, is one of the world's largest multinational corporations and is accountable only to its shareholders. In addition to corporations, celebrities and media figures such as Fearnley-Whittingstall have been crucial in pushing and popularizing eco-labels as a means of resolving the problems of overfishing and rewarding sustainable fishermen.[23] The MSC campaign received backing from major supermarkets in the United States and United Kingdom, such as Walmart and Marks and Spencer, and from powerful NGOs, including OCEAN2012, a coalition of organizations working to transform European fisheries policy. NGOs and civil society actors often represent complex and controversial problems, such as overfishing, through simplistic, one-sided, and populist media campaigns that can play a decisive role in how environmental policies are framed and achieved. Although these agencies and organizations present themselves as equal members of civil society, and are therefore rarely challenged about the campaigns they adopt and encourage, they can have greater access to, and expertise in, financing, media, and governmental decision making than individual fishermen or fishermen's organizations could ever expect to have.

Instead of encountering openness, opportunity, and transparency, many fishermen experience a lack of financial and administrative support and a huge burden understanding what is required to fulfill the criteria for MSC accreditation. In effect, only those sectors of the fishing industry capable of mobilizing considerable organizational, technical, and financial capacities can participate in the "sustainability network" of the MSC. This is reflected in the figures: as of 2011 only three fisheries in developing countries were MSC-accredited, and most of the fisheries currently undergoing certification in developing countries have large-scale industrial features (Ponte and Cheyns 2013). This discrepancy suggests that considerable changes in the organization and profitability of the fishing industry are required before they can hope to productively engage with the MSC accreditation scheme. As Erik Swyngedouw writes, "The new 'gestalt of scale' of governance has undoubtedly given a greater voice and power to some organisations (of a particular kind—i.e., those who accept playing according to the rules set from within the leading elite networks)" (1999, 2003; parenthetical in original).

Recognizing this situation, BIM has now attempted to implement a "transitional" eco-standard for individual fishermen. The "Responsibly Sourced Standard" is designed to "fill the gap" between the EMS and the "gold standard" of MSC accreditation. Rather than being a fisheries-based standard like the MSC, the Responsibly Sourced Standard is an individual operator–based standard that identifies when "catches have been fished responsibly, are of the highest quality and are traceable." Fishermen and onshore operators that qualify for the standard are allowed to display BIM's "Quality Seafood Program" logo on their products. The BIM website specifically states that this standard is intended to enable fishermen to "access the market where fishery-specific certification, such as MSC certification, may not be possible."

The scheme was launched during my time in Castletownbere in collaboration with Frank, a retired fisherman who had committed himself to promoting sustainable and responsible fishing practice. He now works for BIM encouraging the fishing industry to adopt the EMS and to apply for various accreditations. He was recently appointed a director of the Irish South and West Fish Producers Organisation, one of the producers groups that are actively promoting the EMS. When I met Frank, he had just been advertising the label at a Seafood Expo in Paris, where he was trying to get the backing of an international organization. He hoped the label would quickly become known throughout Ireland. Four years later, I was still unable to find the "Responsible Irish Fish" label on products in any Irish supermarkets.[24]

The Green Economy and the New Enclosures

The last chapter examined efforts to reform the top-down, inflexible system of fixed quotas and the problem of discarding. The response from policy makers was to devise a series of new policies and regulations that would channel the activity of individual fishermen toward more efficient forms of fishing and shift the fisheries as a whole toward the measurable goal of MSY. Like the grain producers of the eighteenth century, fishermen were understood as economic agents wanting to maximize productivity and profit in the marketplace. This laissez-faire rationale is, I argued, typical of a liberal economic analysis and gives rise to a particular "regime of truth" in which fishermen are included and excluded according to their ability to adapt to the bioeconomic laws of nature.

In contrast, the EMS and wider system of eco-labeling and accreditation try to achieve environmental and economic objectives by creating a *new* kind of market for sustainably sourced seafood. Individual fishermen are not just incentivized to be more efficient in terms of their fishing activity. They are also encouraged to monitor and assess all aspects of their activity and seek rewards for their performance by actively inserting themselves within international accreditation schemes. The EMS is therefore understood as a way of harnessing individual enterprise in order to respond to the environmental and economic challenges facing the fishing sector. It is clear from the interviews with the BIM representatives that the success of the EMS depends on the "self-empowerment" of individual fishermen; fishermen are expected to take on more responsibility for both their environment and their economic livelihood, and to be enterprising and adaptive within the new "green" economy. This approach is also a response to previous models of bureaucratic, hierarchical state management that have supposedly discouraged individual initiative and failed to offer any realistic strategies for fishermen competing in a global market.

A defining characteristic of this neoliberal form of environmental governance is a shift away from directly regulating the production or extraction of fish—as finite, biophysical resources—to the measurement and valorization of the environmental performance of the fisherman, something that is always open to improvement. The function of the EMS is to document this performance, providing fishermen with a "language," as Jean O'Sullivan said. However, once value is imbued in the appearance of an activity rather than the activity itself, a gap opens up between the situated and complex reality on the ground and the way it is represented through narrowly defined and highly technical criteria. Despite being mobilized through discourses of difference and creativity, the eco-label is thus "an enterprise of normalization in which the particular characteristics of the activities are erased in the homogenization of standards" (Dardot and Laval 2013, 272). The problem, as Mark Fisher recognizes, is not the specific technology of representation but the faith in representation to begin with. The drive to

> assess the importance of workers and to measure forms of labour which, by their nature, are resistant to quantification, has inevitably required additional layers of management and bureaucracy. What

> we have is not a direct comparison of workers' performance or output, but a comparison between the audited representation of that performance and output. *Inevitably, a short-circuiting occurs, and work becomes geared towards the generation and massaging of representations rather than to the official goals of the work itself.* (Fisher 2009, 42; emphasis added)

Although there were no MSC-accredited fisheries operating out of Castletownbere, there was a fish processing and seafood company that had benefited from BIM accreditation, including the West Cork Partnership project described in the introduction to this chapter. West Cork Marine is a local, family-run company based in Castletownbere that employs over one hundred people in its processing factory. The factory is thus of vital economic importance to the town. In addition to providing employment in the factory, it processes and exports a variety of shellfish products supplied from inshore boats around the southwest coast, including from five of the company's own boats. The bulk of its whitefish, however, is imported from Iceland. The company collects shipments of whitefish at Cork airport once a week, drives them down to Castletownbere, processes the fish, and then redistributes them to national and international retailers. Many fishermen who are opposed to the processing of imported fish organized a protest at Cork airport several years ago, but others see it as the future. "We have to get smarter," the owner of a trawler told me. "They [the company] get their fish from Iceland because it's the same price as from the co-op but always consistent, packaged, and certain. If you asked the co-op for fifteen kilograms of plaice, they'd laugh at you." He knew that only those who were "proactive" would survive: "We have to do something. We're at a standstill: it is survival of the fittest." Shipments of fish from Iceland are more consistent, allowing the company to supply to supermarkets and other large retailers that demand regular, reliable supplies of fish. For similar reasons, the company set up another processing factory in Indonesia because it is cheaper to process prawns there.

West Cork Marine has won several awards from BIM and secured funding on the basis of its economic innovation and environmental credentials. It received thirty thousand euros from the LEADER program to "develop its brand." The company employs seven people in its marketing division. In the supermarkets around Ireland, its products are clearly advertised on the basis of their provenance: they come from the "clean waters of the South West of Ireland." But even this marketing drive is not enough.[25] Responding to growing competition and expanding markets, the Irish government requires that companies like this one in Castletownbere capture more of the market by becoming even more responsive to consumer demands. West Cork Marine cannot depend on the fragmented, inconsistent catch of the Castletownbere fishing fleet, nor can it rely on its existing distribution network and retailers. A market-orientated strategy means increasing vertical integration, allowing the company to streamline its activity and connect the point of production to the point of consumption ever more seamlessly. In a "network" society, as Andrew Barry puts it, the successful actor is the one "who has the time to keep continually informed and in touch, and who is able to communicate and travel anywhere for his business. He is not rooted in a place, but in a web of extended connections; able to draw things together without ever forming them into a vision of the whole and without having need to do so" (2001, 13).

A market-oriented fishing industry competing on "green" credentials involves shifting value away from the activity of fishing itself and toward market competition based on the availability of information and the manipulation of appearances. "You can't fish well and do the other things on land," I was told by several fishermen. Those who continue to fish because that is what they know and do are unlikely to be able to adapt to the new realities of the seafood industry. "I have a family. I'm up to my fucking eyeballs," one fisherman told me. "I don't have time to go away chasing this [eco-labeling] and that is the fucking problem."

The two men who own the company in Castletownbere were once fishermen. They recognized early on, as many now do, that there was no money to be made in fishing alone and that the future was in marketing. In this way, a "local" company such as West Cork Marine is part of a new configuration that appears to deliver on the values of cooperation, differentiation, and environmental and community value while profiting from the "old" values of homogenization and profit-maximization. The disparity between the representation of sustainability and the social and material reality underlying it was clear to anyone who cared to look; everyone I met in Castletownbere understood the "success" of West Cork Marine as a consequence of "playing the game." Although many respected what the two owners had done and were grateful for their employment, both directly in the factory and through the routes to market provided for inshore fishermen, there was also a recognition that the success came at a price, an attitude perhaps best summed up by a comment made about the contributions this local company gave to the local Royal National Lifeboat Institute (RNLI): "They are generous but that money was taken from somewhere else."

The EMS represents a dovetailing of environmental and economic objectives: fishermen are located between the ethical obligation to "care" for their environment and the need to "innovate" and compete with other fishermen around the world. Underlying the expansion of individual responsibility is the supposed expansion of individual freedom, the growth of choices and options that are intended to enable individuals to take more control over their lives. But the process of securing accreditation and becoming "visible" as a responsible "steward" is not transparent or equitable: "The supposed 'self-empowerment' of this system rests upon a simultaneous imposition of external control from above and internalisation of new norms so that individuals can continuously improve themselves" (Shore and Wright 2000, 61). Once again the unfolding narrative of ecological modernization involves the cultivation of a particular type of economic subject, but in the neoliberal guise this means "the implantation of particular modes of calculation into agents, the supplanting of certain norms, such as those of service, and dedication, by others, such as those of competition, quality and customer demand" (Rose 2006, 153). These accreditation schemes encourage and reward a different type of fisherman, one who is not simply efficient and productive in terms of marine resources but also skilled in the manipulation of information and enterprising in how he devises ways to access emerging transnational "sustainable" commodity markets and governance networks.

For many of the fishermen in Castletownbere, however, the vision of ecological modernization embodied in the EMS and eco-label involves a loss of autonomy, a growing dependency on global regulatory authorities and markets, and a disconnection from the local, social, and material contexts in which they continue to eke out a livelihood. For these reasons, labeling schemes and the EMS have not been supported by everyone working in the fisheries sector. When I asked Maeve O'Reilly, who also worked for BIM as a fisheries manager, about the benefit of the EMS and the MSC accreditation, she acknowledged they could help fishermen get a better price for their catch, but she said this did not necessarily help with the more pressing challenge of managing the actual fish stocks: "So you do an EMS, I get rid of my plastic in the proper way. I adhere to the minimum landing size. Well bully. That's just what that is. It has nothing to do with the fishery being biologically or economically viable" (O'Reilly 2009). She picked up a publicity leaflet sponsored by BIM. Its advertisement for Kerry lobsters read, "Kerry lobster: Fit for a King." The leaflet included pictures of fishermen in beautiful surroundings and expressions such as "commitment to sustainability" and "pure waters." She said its promises were meaningless because there were still no management tools in place to limit the amount of fishermen who could catch lobster or the amount of pots a fisherman could drop. It is out of this perceived need to re-ground fisheries management within specific biological and economic contexts and the need to limit fishing effort that a third form of liberal governmentality emerges.

4

Community-Managed Resources

A "Third Way" for Environmental Governance

Beyond the State, Beyond the Market

Most of the boats I went out on during my research were the small, inshore boats that head out early in the morning and return in the early evening. They are usually between eight and twelve meters long and do not carry the huge nets required to trawl for finfish. For the most part, these boats are operated by their owners and one other crewman. Although some of the boats in the inshore sector have polyvalent licenses—meaning they can catch both finfish and shellfish—most of the inshore boats target and rely on shellfish, primarily lobster, brown crab, prawns, and scallops.

Technically, the inshore sector describes a territorial area rather than a type of fishing or boat size; the inshore extends out six nautical miles from the national coast of each European member state, an area that does not fall within the jurisdiction of the European Common Fisheries Policy.[1] Historically, the inshore fisheries have been the social and economic mainstay of small coastal communities: in Ireland the inshore sector accounts for about half of all those employed in the fishing industry and over three quarters of the vessels in the fishing fleet (Bord Iascaigh Mhara 1999). Compared to the offshore fishing industry, the inshore sector has not experienced rapid technological advances or overcapitalization in terms of expanded fishing capacity. As a consequence, the inshore sector is often described in policy documents as "traditional" or "artisanal." One of the consequences of the social and economic composition of the inshore sector is that while inshore fish stocks have declined over the past few decades, they are not under the same kind of pressure as offshore fish stocks. Despite the lack of significant national or European regulation, the inshore sector has therefore managed to remain relatively sustainable.[2]

Beginning in the early 1990s, however, fishermen began seeing declines in the catch levels of lobster, the main commercial species in the inshore fishery. Several groups of inshore fishermen, mostly from the south coast, pushed the Irish government to introduce statutory regulations on fishing and some form of community-managed lobster fishery framework. In response, the Irish government began developing a management plan for the Irish inshore fisheries. This plan immediately ran into difficulties because the inshore fisheries consist of many unlicensed part-time fishermen. After a long, drawn-out process, BIM finally published a consultation document for the lobster fisheries in 2008. The document proposed introducing sixteen territorial units around the Irish coast as a basis for managing lobster stocks. Only licensed fishermen would be able to fish the stocks within their area. In theory, this improved security of access and control would allow communities of fishermen to work together to achieve specific objectives, including conserving lobster stocks and improving the marketability of their catch.

Efforts to implement a community-managed model in the Irish lobster fisheries can be understood within a general turn within environmental policy-making toward more participative, community-based resource management. Starting in the 1980s, a body of academic research responded to the perceived failures and weaknesses of existing models of resource management. Put simply, these models proposed a limited choice between some form of privatization (as with the ITQs) or top-down, centralized state ownership and control. The work of economists and anthropologists complicated this picture by documenting and analyzing how communities around the world have developed formal and informal ways of managing and allocating resources themselves (see Hanna et al. 1996; Ostrom et al. 2002; McCay 1990). Crucially, these often complex and adaptive systems of resource access and distribution were generated by the users themselves and not through the external authority of the state (Ostrom 2008a).

Throughout the 1990s and 2000s, models of community-based resource management became an increasingly important policy instrument for the management of natural resources, including fisheries (Agrawal 2003; Leach 2008; Li 2005, 2006). Community-based approaches grew in popularity because of a belief that they better reflected the specific social, economic, and environmental needs and characteristics of local communities while also fulfilling national (and supranational) economic and regulatory objectives. Ideally, this approach promises greater democratic participation, more sustainable forms of resource management, and more equitable distribution of and access to resources (Baland and Platteau 1996; Berkes and Folke 1998). This "third way" for resource management has even received conditional support at the highest levels of international governance. As early as 1992, the World Bank stated, "Governments need to recognise that smaller organizational units, such as villages or pastoral associations, *are better equipped to manage their own resources than are large authorities and may be a more effective basis for rural development and rational resource management than institutions imposed from the outside*" (World Bank 1992, 143; emphasis added).

During my research, I interviewed several fisheries managers and policy makers involved in efforts to implement a form of community-based management for the Irish lobster fishery. Just as Jean O'Sullivan had been influenced by the development of environmental auditing systems in Australia, these resource managers knew about successful models of community-managed fisheries from around the world, particularly the Maine lobster fishery, which remains one of the success stories of the community-based approach. The managers I spoke to had all worked in the fisheries for a decade or more. They were acutely aware of the limitations of top-down systems of management, particularly in the context of the highly fragmented, diverse inshore fisheries. This is a familiar argument in the literature on small-scale fisheries management, where complex ecosystems in remote areas make it costly and difficult for more centralized forms of management to operate (Jentoft 1996; Jentoft and McCay 1995; Kooiman et al. 2005; Kompas and Grafton 2004; Berkes and Pomeroy 1997). Similarly, they were critical of the individualized, market-based measures discussed in the previous two chapters—both the ITQs and the EMS and MSC accreditation schemes. While these were identified as potentially useful and important measures, they were understood as being of limited value when dealing with the social, economic, and environmental composition of the inshore fisheries.

In this chapter, I argue that the third way for fisheries management—community-based resource management—represents another variant of the liberal approach to managing the problem of overfishing. What makes the community-based approach novel is that it seeks to create situations in which individuals are incentivized to work *together* to achieve policy goals. This approach involves the delimiting of new boundaries of access and inclusion that do not rely on the traditional categories of private property or direct state control. These "softer" boundaries are still justified as necessary interventions to channel individual self-interest toward the more "efficient" exploitation of limited marine resources. Although this approach undoubtedly represents a departure from the traditional public/private binary, this does not mean that the state or the market has "retreated" as a new collective form of resource management has advanced. This new form of environmental governance extends familiar bioeconomic rationalities through new (local) scales and (collective) subjects of action. By reconfiguring relations between the state, market, and individuals, this mode of governance carves out new spheres of freedom and participation, while at the same time excluding those who might operate according to different social and ecological values and ways of knowing (Mosse 1997, 471).

This chapter provides some background on the Irish lobster fisheries and the recent development of the community-based management plan. While support for better regulation of the lobster fisheries initially grew from fishermen themselves (with support from some fisheries scientists), it is important to situate the subsequent response from the Irish government within a broader series of arguments supporting the role of communities in the management of local resources. Government reforms responded to a sustained critique of both centralized state management and privatization in the area of resource management. The successful implementation of one of the best-known and frequently quoted examples of community-managed fisheries—the Maine lobster fisheries—in 1995 and the historic challenges of managing the Irish lobster fisheries ensured that there was considerable hope and promise attached to such models in Ireland.

After addressing the origins of community-based fisheries management, I examine how the management plan for the lobster fisheries and the scholarly literature on "common-pool resource" (CPR) management share a liberal understanding of the problem of resource depletion and subsequently apply economic reasoning to resolve it. Rather than challenging the core liberal assumptions of Garrett Hardin's "Tragedy of the Commons" thesis, community-based approaches tend to reproduce them by assuming that fishermen are individual, economic actors exploiting common natural resources in a situation of open access. Where this perspective differs from Hardin is in identifying ways of governing individual behavior that do not rely on direct regulation or privatatization. Fisheries managers made it clear to me that the community-based approach to the lobster fisheries is about developing more *effective* forms of governance in order to achieve the same bioeconomic goals of sustainability. While appearing pragmatic and neutral, the third way for resource management produces its own "regime of truth," a new set of bioeconomic parameters determining which actions are "environmental" and which ones are not.

The Promise of Community-Based Resource Management

The most valuable inshore species in terms of earnings is the lobster. The lobster fishery overlaps with the crab fishery because both species are caught at the same time of the year using similar technologies and techniques. Depending on the weather, the season runs from March/April to October/November. To catch lobster, fishermen use "pots" that were traditionally made of wood but are now mostly made of wire. These pots are essentially small cages with two funnel-shaped openings on either side that allow the lobster to enter the pot easily but not get out. Fishermen fill the pots with partially rotten mackerel or other fish as bait for lobster, and fresh fish for crab. They string together thirty to fifty pots on a rope or "string," "shoot" them out at sea, and leave them on the sea floor for two to four days. Buoys marking the locations of the strings are a common sight all along the coast during the season. A fisherman might have close to a thousand pots in the water at any one time and check five hundred every two or three days depending on the time of year and the weather.

In the nineteenth century, lobster fishing was a part-time occupation of Irish fishermen-farmers (Mac Laughlin 2010). Although there were an estimated 5,965 boats and more than 23,000 fishermen harvesting lobster in 1876, limited gear technology and a mostly subsistence form of production meant that stocks were not overexploited (Browne et al. 2001). Improved transportation for shellfish and the opening of British and European markets led to the development of the lobster industry in the late nineteenth and early twentieth centuries. Between 1900 and 1912, lobster landings went from 128 to 228 tonnes. By 1927 the annual catch had grown to 430 tonnes. Catches declined up until the Second World War, rose to a peak of 350 tonnes in 1959, and then stabilized until 1967. There are no records of landings between 1967 and 1994, but in 1994 boats landed 715 tonnes of lobster. The catch fell to 513 tonnes in 1997 before landings reached a record high of 853 tonnes in 2004 (Tully et al. 2006).

Up until 1992, the only regulations on the lobster fishery were restrictions on young lobster—a minimum landing size of 240 millimeters (total length) or 85 millimeters (carapace length)—and a ban on the capture of lobsters by SCUBA diving. Lobster fishermen were also supposed to have a license, but many part-time fishermen did not see the need to recognize even this minimal provision. Regulations began to change in the early 1990s. Ronan Browne and others argue that the catalyst for change was pressure from concerned lobster fishermen, who first observed a declining catch per unit effort (CPUE)—fewer lobster in the same number of pots. This decline led to the establishment of lobster hatcheries in 1992 and 1994 in Galway and Wexford, where the organization of lobster fishermen was strongest. The hatcheries, which cultivate lobster larvae from wild female hen lobsters and then release them when they reach maturity, helped mobilize local fishermen. "These hatcheries," Browne writes, "acted as focal points for seminars, conferences and information exchange. Education and debate were also provided at local meetings around the coast by industry and scientific representatives from the USA (Maine) and Ireland" (Browne et al. 2001, 51).

Led by the South Wexford Lobster Fishermen's Co-op, the voluntary conservation practice of "V-notching" began on a limited scale in 1993. "V-notching" is a fishing practice that involves marking mature female lobsters with a small "v" in the tail. The South Wexford initiative encouraged fishermen to return marked lobsters to the sea. The program was supported by BIM with the help of European and national funding (Browne et al. 2001). The following year legislation to protect "V-notched" lobsters was introduced. For some industry members, scientists, and fisheries managers, however, these minimal regulations (landing size, "v-notching," and SCUBA restrictions) did not go far enough to ensure well-managed, sustainable lobster fisheries.

In 1994 regional lobster fishing associations and co-operatives were formed around Ireland under the umbrella of the Irish Lobster Association (ILA). In Castletownbere, I got to know one of the leading figures in the ILA, a fisherman named Tom. He told me about a visit he and other representatives made to Maine in 1994. They had been invited by the Maine Lobster Association (MLA), which was successfully moving at that time toward a form of co-managed lobster fisheries that would ensure that the lobster fishermen had direct control over their fishing activity. The trip was an important moment in the evolution of community-managed resources as a new field and practice of policy-making.

In 1995, the year after Irish lobster fishermen visited Maine, the Maine State Legislature passed the "Zone Management Law," formally recognizing the role that lobster fishermen played in managing their own fisheries. The law established a limit of twelve hundred traps by the year 2000, a trap tag system to identify owners of traps, and an apprenticeship program for new entrants into the lobster fishery. The most significant and innovative aspect of the law was that it gave the lobster fishermen control over key aspects of the lobster fisheries. Lobster license-holders elected councils to manage separate coastal zones. Each zone council was empowered to set limits on the number of traps each license-holder could fish, the number of traps that could be fished on a single string, and the time of day when fishing was allowed (Acheson 2003). Two-thirds of the license-holders in a zone had to approve of any new measures before the state commissioner of marine resources could establish them as departmental regulations and enforceable regulations by the marine police. Although there was evidently some doubt about how well this system of self-regulation would work, the comanagement law was soon considered a success: by 1998 all seven of the zones had passed trap limits, and by 2001 five of the seven zones had established limited-entry rules for their zones. This successful example of comanagement has now become one of the benchmarks for comanaged fisheries around the world (Caffentzis 2008).

These developments in policy making occurred alongside growing scholarly publications and debates about the question of resource management (Agrawal 2003). Anthropologist James M. Acheson, a leading scholar on community-based management and an important mediator in the development of the Maine "Zone Management Law," describes the late 1980s and early 1990s as a period when "many people interested in fisheries management had become convinced that *standard ways of managing fisheries were failing, and that new approaches needed to be tried*" (Acheson 2003, 99; emphasis added). Anthropologists were finding diverse examples of communities around the world that continued to manage natural resources, such as fisheries, forests, and water, through collective forms of decision making and allocation (Berkes 1987). Alongside this rich empirical work, political scientists, economists, and sociologists tested the assumptions of rational choice theory by showing through laboratory tests and simulations that individuals could act in concert to resolve resource dilemmas (McCay 1990, 1996; Hanna et al. 1996; Ostrom and Schlager 1992; Ostrom 2000a). Instead of always acting out of self-interest, their subjects found ways of cooperating to resolve shared problems to the benefit of all.

In the introduction to her seminal text *Governing the Commons*, environmental economist Elinor Ostrom makes it clear where the motivation for researching the rules and institutions of collective action lies: the power of certain dominant metaphors for explaining the causes of resource degradation, which results in a limited choice of management strategies. Although Ostrom does not cite Malthus's parable of the "mighty feast," she does single out Hardin's "tragedy of the commons" as a key example of such explanatory narratives. In his 1968 essay, Hardin describes an imaginary situation in which a group of herdsmen graze their animals on a common pasture. Because the pasture is open to all, he argues, each herdsman is able to increase the number of animals he grazes without any restriction while the cost of the increase is distributed among all the other users of the pasture. The pasture is ultimately destroyed by overgrazing: "Therein lies the tragedy. Each man is locked into a system that compels him to increase his herd without limit—in a world that is limited. Ruin is the destination toward which all men rush, each pursuing his own best interest in a society that believes in the freedom of the commons" (Hardin 1968, 1244).

Ostrom shows how this simple story recurs as the basic, commonsense explanation for resource degradation. Not surprisingly, she quotes numerous examples from the fisheries, including the analysis of H. Scott Gordon, a fisheries economist. Over a decade before Hardin, Gordon wrote:

> There appears . . . to be some truth in the conservative dictum *that everybody's property is nobody's property*. Wealth that is free for all is valued by one because he who is foolhardy enough to wait for its proper time of use will only find that it has been taken by another. . . . The fish in the sea are valueless to the fisherman, because there is

no assurance that they will be there for him tomorrow if they are left behind today. (Gordon 1954, 124; emphasis added)

To reiterate her point, Ostrom quotes a statement from the Canadian minister of fisheries and oceans in 1980: "If you let loose that kind of economic self-interest in fisheries, with everybody fishing as he wants, taking from a resource that belongs to no individual, you end up destroying your neighbor and yourself. In free fisheries, good times create bad times, attracting more and more boats to chase fewer and fewer fish, producing less and less money to divide among more and more people" (Ostrom 2008a, 8).

The familiar interpretation of Hardin's thesis is that the only solution to the "problem" of the commons is to enclose the resource through the allocation of private property rights or direct state ownership (Dietz et al. 2003). We have already seen how a system of tradable quotas is currently being implemented in the European offshore fisheries in response to such a "vicious cycle" of overexploitation. For Ostrom and her colleagues, however, this conclusion is based on the inaccurate portrayal of resource users as "helpless individuals caught into an inexorable process of destroying their own resources" (Ostrom 2008a, 8). Ostrom's great contribution to the question of how we organize biophysical resources and systems is in challenging this representation and complicating the story. Rather than being helpless in the face of resource problems, individuals in very different parts of the world have come together to devise and implement collectively binding and beneficial rules that have successfully managed and sustained common-pool resources.[3] Their action challenges not only the belief that all individuals are by nature self-interested and separated from one another but also the idea that some form of external authority is necessary to govern these communities. The challenge for resource managers is to identify what principles and conditions are favorable for producing such institutions of collective action. As Ostrom concludes, "I would rather address the question of how to enhance the capabilities of those involved to change the constraining rules of the game to lead to outcomes other than remorseless tragedies" (2008a, 7).

Tom, the Castletownbere lobster fisherman, told me that the visit to Maine had had an enormous effect on him and the other fishermen who had made the trip. He described how his American counterparts were now catching many more lobsters in their pots compared to before, and that this was due to the greater control they had over their own fisheries. Not surprisingly, when he and the others returned to Ireland, they redoubled their efforts to legalize community-managed lobster fisheries. In addition to compulsory "v-notching," they lobbied for a minimum landing size for lobsters and, most significantly, a form of managed access to the fishery that would enable groups of local lobster fishermen to regulate themselves and the lobster stocks they relied on.

Despite these efforts, there would be no visible effort to regulate access to the lobster fishery until 1998. In that year, the minister for the marine and natural resources, Dr. Michael Woods, commissioned BIM to carry out a "strategic review of the inshore fisheries sector." His request was partly a response to the lobbying efforts of the lobster fishermen's organizations and partly a recognition that some form of management was necessary to prevent the continuing decline in lobster catches, which are so crucial to the viability of many small coastal communities. It was the first time any specific strategy for the inshore sector had been attempted. While the inshore sector makes up the majority of vessels in the entire fishing fleet and employs nearly half the fishermen, it only provides a fraction of the industry's overall income and does not fall within the European CFP.[4]

In May 1999 BIM released a watershed paper, "Irish Inshore Fisheries Sector, Review and Recommendations." It made the inshore sector part of the government's overall policy for the fisheries sector. The report proposed that a national Inshore Fisheries Advisory Committee (IFAC) be established as a forum for discussing "policy and strategies for the inshore sector and to act as a consultative/liaison forum." The committee included representatives of BIM; the Department of the Marine and Natural Resources; the Marine Institute; Central and Regional Fisheries Boards; and representatives of the fishing industry.

It was not until 2008, however, more than a decade after Tom's visit to Maine, that BIM published an actual consultation document, "Managing Access to the Irish Lobster Fishery." It provided the basic outline for the projected management of the lobster fishery and offered a blueprint for the management of other shellfish species. The management plan closely followed the Maine lobster fishery system. To begin with, the Irish coast was to be divided into eight zones or territories. Each of these zones was supposed to be broken into two additional units to allow more flexibility for those fishermen who were working on the boundary between two units.[5] While the Irish lobster fishery has nothing like the well-developed territories and "gangs" of the Maine lobster fishery, the decision to create localized zones was based on the need to create a sense of ownership among small communities of fishermen over particular lobster stocks. The boundaries of the zones were therefore chosen partly through consultation with fishermen and partly through existing knowledge about the geographical limits of lobster stocks.

The significant aspect of the proposed management plan is that for the first time a form of regulated access to the lobster fishery would be legislated for and enforced by the state. Fishermen who currently had a license to fish for lobster would be issued an authorization that would allow them to fish for lobster within their zone. There was a limit placed on the number of authorizations issued. These authorizations could not be bought or sold and thus did not function as a tradable right of access. There were no criteria set for how many pots a fisherman could set, but the hope and expectation was that the security promised by the limited entry for fishermen and the scale of the fishing zones would foster community collaboration and collective decision making in the management of fishing practices.

Hardin Revisited: A New Solution to an Old Problem

Three people involved in the design and implementation of the community-based strategy for managing the lobster fisheries were James Carney, a scientist with the Marine Institute; Maeve O'Reilly, a BIM fisheries manager responsible for the inshore fisheries; and

Brid Smith, a senior civil servant in the Department of Food, Agriculture and Fisheries. They all emphasized that the rights of access being proposed for the lobster fisheries were not private rights that could be bought, sold, or leased by individual fishermen. The comanagement of the lobster fishery, they all asserted, was deliberately being designed to *avoid* the social and economic costs of privatization. O'Reilly and Carney both quoted the examples of the Tasmanian rock lobster fishery, where ITQ had created monopolies within fisheries that had previously supplied an "expanded social benefit."

> *The one thing we're saying with the lobster is you can have an exclusive right, but not a transferable right—that's the difference with ITQ.* That will enable the long-term view without creating this great cost for the next generation who want to get into the fishery. If you read the ITQ literature, it's always split fifty/fifty down the middle because eventually it comes down to people's politics. That's what it is at the moment: our politics, the government's, is to avoid ITQs. (Carney 2009; emphasis added)

What is considered "political" from this perspective is the *transferable* nature of access rights. Everyone agrees on the need to regulate access to the fisheries, limit fishing activity, and provide greater certainty for the fishermen, but the question is how this should be achieved. Carney told me, "Well I think if I was a lobster fisherman at the moment, irrespective of how much I want to see the long term, I don't think I could act in the long term. I couldn't be proactive because the management system doesn't allow me to. It effectively forces me into a short-term view. So that has to change certainly" (2009). O'Reilly put it best, articulating the need to balance the economic motivations of individual fishermen with the common interest:

> If I was a punter with a license, and particularly if I was in a certain economic position where I had to provide for my family, then of course I would be wanting what made me the most money and gave me the most security. *So there's an element of fighting against the individual good for the common good that is difficult and there is a risk if it's not*

done properly that we won't succeed in ensuring a common good into the future that doesn't end up in massive aggregation. And aggregation has been addressed and discussed in terms of how you would circumvent that without being totally biased against enterprise. (O'Reilly 2009)

The decision to establish territorial units for the lobster fisheries and manage access to these units stems from an explicit need to accommodate individual self-interest within any management framework. This formulation of the problem of overfishing does not, in other words, deviate from Hardin's "tragedy" or liberal economic accounts more generally. The assumption that overfishing is caused by the unregulated economic activity of individual fishermen, and the forms of analysis that emerge from this assumption, are captured in this quote from Susan Hanna, a marine economist and outspoken advocate of community-managed resources:

> When a fishery is open to anyone, there is no assurance that a fish not caught today will be around tomorrow. In fact, it will probably be caught by someone else. So, why not catch it yourself? Why invest in the long-term sustainability of the fishery if what happens tomorrow or next week or next year is highly uncertain? *It's not rational. . . . The uncertainty among fishermen about whether fish will be there tomorrow interacts with the economics of fishing and the natural variability of fish populations in a way that is destructive of the resource. Attempts by the government to manage fishing behavior are inadequate to control these dynamics.* (Hanna 1999, ix; emphasis added)

Hanna is not here describing the "irrationality" of Hardin's "commons" but rather the irrationality of "open access." This is a key distinction for advocates of community-managed resources. As Ostrom writes, the commons describes a situation "where the members of a clearly demarked group have a legal right to exclude nonmembers of that group from using a resource. Open access regimes (*res nullius*)—including the classic cases of the open seas and the atmosphere—have long been considered in legal doctrine as involving no limits on who is authorized to use a resource" (2000b: 335–36). This distinction opens up space for collective forms of resource management while at the same time continuing to assume that the subject of these collective institutions and norms is the self-interested, economic subject. This analysis perpetuates the "methodological individualism, self-interested rationality, rule guiding behavior and maximizing strategies" that are most often associated with Hardin and liberal economic perspectives on resource management (McCann 2004, 7; Mansfield 2007a). In other words, the third way for resource management does not deny the "tragedy" scenario but suggests that there are better ways of resolving it beyond direct state intervention or private property. The institutions of the commons are no more than alternative rules and norms for shaping the same individual subject who conducts their activity through the application of economic reasoning (Forsyth and Johnson 2014). Carney clarified this perspective:

> The setting up of management units, limited access, and managed entry to those units is the first basic step to address the sustainability issue. *Now you have managed access, when you have managed access the group can collectively act.* They may wish to reduce effort at a certain rate, to improve the cost benefit. So it gives them scope for more action essentially. . . . The question is then why will they develop efficient collective action now if they haven't in the past? *In my opinion it's because of the additional—I don't want to call it property right, but the additional, approved tenure-ship and security they are given.* And we have been very strong in trying to put forward that point: this is not a threat, it is a provision, it is giving you something, a better exclusive right to the resource. (Carney 2009; emphasis added)

Carney is unable to specify the type of property right that the new system of managed access provides, but he recognizes that its *effect* is the same as other property rights in terms of providing the individual and the community with a form of security that will encourage rational action by individuals and collectives. The hesitation and inconsistency with which Carney, Smith, and O'Reilly spoke of "property," "tenure," "ownership," and "security" illustrate something of the real-world complexity that comanagement strategies attempt to address. The key point for resource managers is that some form of limited access can be implemented that shapes and channels the actions of individual resource users toward the desired environmental and economic goals. How this is done is less important than the outcome (Dietz *et al.* 2003): "Our main goal," Ostrom writes, "has been to show . . . that the techniques of rules and games work and put policy analysis on a sound logical footing" (Ostrom et al. 1994, 96–97).

This pragmatism was already articulated in an essay written in 1992 by Elinor Ostrom and Elena Schlager. In the essay, the two authors discuss the need for an expanded concept of property rights. They define property-rights institutions as "sets of rules that define access, use, exclusion, management, monitoring, sanctioning, and arbitration behavior of users with respect to specific resources" (Ostrom and Schlager 1996). Ostrom and Schalger argue that the supposed choice between private and public property-rights regimes is too simplistic and does not acknowledge the diversity of property-rights institutions that operate in real-world contexts. Rather than being a clear, precise choice between private or public ownership, studies of many local systems of resource use suggest the existence of "diverse bundles of rights that may be held by the users of a resource system" (Ostrom and Schlager 1992, 249). Although the same liberal problematic applies, the innovation of community-based resource management is in opening up new possibilities for policy makers and resource managers keen to deliver more flexible, adaptive, and effective institutions for resource use and management (Auer 2014): "No real-world institution can win in a contest against idealized institutions," Ostrom and Schlager write. "The valid question is how various types of institutional arrangements *perform comparatively when confronted with similarly difficult environments*" (Ostrom and Schlager 1992, 260; emphasis added). The challenge is to get the "institutions right" (Ostrom 2008b). This reasoning becomes apparent in the context of the lobster management framework.

One of the main reasons there has been no regulatory framework in place for the Irish inshore sector is the extent to which the fishermen and fishing activity are so geographically dispersed and irregular. There are large numbers of unlicensed fishermen who require very little gear to conduct inshore fishing for crab or lobster. In Castletownbere, for example, I lived above a small natural harbor where two or three boats were tied up for fishing during the summer months. It was easy for the owners of these boats to lay a few strings of pots without anyone stopping them or even knowing what they were doing. Attempts to manage this situation directly would be far too costly and ineffective, an argument that applies to small-scale inshore fisheries around the world (Acheson 2003). Unlike Maine, Ireland has no well-established community organizations or "gangs" that operate within identifiable territories or fisheries. A common remark made about the lobster fishermen was the extent of their "fragmentation" and "reluctance to participate." As Bill Whelan, a BIM fisheries manager with responsibility for the management of the lobster fishery, told me, "the inshore fisheries, the people involved in it, are very disparate. They're not united, they do not have articulate representative organizations like the Producer Organizations do" (2009). While co-ops had formed in the early 1990s, they had never been strong. They did not represent the majority of fishermen operating along the Irish coast, and they lasted no longer than a decade. This is related to the limited economic profitability of the Irish lobster fisheries.[6] Unlike the Maine fishermen who specialize in lobster fishing and can make significant incomes, Irish inshore fishermen continue to target different species, including crab, prawns, and finfish during the winter.

In the late 1990s BIM and the government had to identify the numbers of fishermen and vessels operating in the inshore sector. They soon discovered that half the boats operating commercially were not even licensed. Many of the smaller fishermen had not bothered to sign up because, as Brid Smith told me, "the idea of bothering to license his boat was just mad" (2009). Many of these boats were just small punts with an outboard motor, and anyone who lived by the coast and wanted to catch a few crab or lobsters on the side could do so. Having identified a large number of unlicensed fishermen, the government then had to decide what to do with them. In 2002 the European Commission announced that no new boats could be added to the Irish fishing fleet. In order to circumvent this limit, the Irish government decided to grant the unlicensed boats a free potting license. It entitled them to fish for shellfish but no other species. In general, the potting licenses went to smaller operators who were not full-time fishermen. They caused some tension because established full-time fishermen saw the extension of these free licenses as a threat to their livelihoods. Between 2006 and 2012, when the expanded licensing process took place, the number of vessels in the shellfish fleet increased by 64 percent. Despite the tension this caused, the licensing of all active fishing boats was understood to be a necessary first step toward the implementation of a management structure for the inshore fishery. Smith, who had worked in the fisheries since the late 1990s, recalled,

> A huge effort went in to bringing them into the fold. The plan was effectively set back to get this done, and after that we then started: okay, these are the boats, the inshore boats, these are the boats that target lobster say. What are we going to do to manage them? Again the message is hasten slowly, no big steps, no mad stuff, because we could have the most gorgeous management framework in here in Clonakilty or in Dublin but it wouldn't be worth the paper it was written on. *You know piles of paper supporting it but for anyone who looked beneath it it would have no impact on the ground.* (Smith 2009; emphasis added)

During her time in the industry, Brid Smith had been strongly involved in the development of a management policy for the inshore sector. One of the reasons it had taken so long to evolve, she told me, was this commitment to moving slowly to get things right "on the ground":

> First of all the principle behind all of this, and we've gone out to international experience, we've had people in from Canada and New Zealand, other places that have also tried to grapple with this, and the basic principle is the nature of inshore fishing: management by centre is not an option. It just doesn't work. It's way too dispersed. The idea that you'll have a police man or inspector standing out in every cove

> checking out whose fishing and who is not fishing is just mad. No way. Even in the good times when there was money for public servants it wasn't an option. *International experience tells us bottom up, bottom up, bottom up and not to force the process, just keep going back and back until you are driven demented by the whole process.* Back down to them, back down to them, make sure they have ownership and buy in. Ownership and buy in are absolutely key. (Smith 2009; emphasis added)

Carney, who has been providing data and policy recommendations on the Irish inshore fisheries for over two decades, emphasized the need not only for community "buy-in" but also for different kinds of knowledge that could account for the complexity of the inshore sector and the marine environment. He made it clear that perfect knowledge of the marine environment would never be possible and a management system based on this premise would never work. Any new management system instead had to incorporate an element of flexibility and responsiveness that would reflect the biological, social, and economic complexities of the lobster fishery. He told me how the science and management of the lobster fisheries had departed from past practices and why as a result the active participation of lobster fishermen in the gathering of data and in managing the fish stocks was vital to this transformation:

> In the case of lobster, we haven't gone down the road of trying to assess population abundance really, or even in making forecasts about abundance, or future catches and future population sizes. It's not possible to do it really is what we feel, and no one in Europe is doing it really and we are mindful of the difficulties that assessments say of whitefish and pelagic in ICES are getting into, they're getting into difficulty because of poor data quality, or because it's just not possible to make these predictions. *So there's a move away from using those predictive models to using what we would call just simple indicators of the current status of the stocks. So rather than having an absolute estimation of something, you have a relative indicator. So you're looking for trends over a period of time rather than absolute point estimates at particular points in time.* So they are different scientific approaches. *The scientific approach and the certainty with which scientists can make an*

> *estimate is also related to, and in some way defines, the way the management system works.* (Carney 2009; emphasis added)

Echoing the views of other fisheries scientists, Carney emphasizes how the uncertainty and complexity of the marine environment changes the role of science, migrating away from formal truth claims to a more grounded, pragmatic definition of "truth" within a prescribed consensus on resource conservation. Scientists no longer provide precise figures from which fixed quotas and other limits can be determined and implemented. Instead, they aim to provide ongoing assessment of the overall health of particular fisheries using a range of indicators that can monitor the entire life cycle of the lobster and its interactions with the fishing effort and strategies of the fishermen (Berkes and Folke 1998; Freire and Garcia-Alut 2000). "Relative indicators" may not be precise, but they trace patterns and tendencies over time that allow fisheries scientists, managers, and fishermen themselves to identify whether certain actions and practices effectively sustain the biological health of the fishery or not. These indicators relate to stable biological processes such as growth potential, migration, growth rates, and habitat in addition to economic activity such as catch per unit, price, and the overall number of pots in operation. Carney summarizes how this kind of approach to the nature of bioeconomic processes transforms the way resources need to be managed:

> If we were absolute and precise about everything, then we could give an absolute and precise answer [and] then there's no debate and therefore a top down model could work. . . . *But because we don't have that we need consensus and you need to build the management process and the system so it can account for, or cater for the uncertainty and unpredictability, and can adapt quickly if need be.* And hence the move toward the more comanagement model rather than a top-down situation. *So I think it's important that the scientific work which is traditionally seen as the starting point is now seen as one seat at the table, and can give its opinion, and is only one opinion.* That is where we are at the moment I think. (Carney 2009; emphasis added)

While the objectives of fisheries management remain the same in terms of the conservation of fish stocks and the economic viability of the fishermen, the form of governance required to achieve these goals in the inshore fisheries necessarily differs from the regulation of the offshore fisheries. This is because the characteristics of the inshore sector are *different* in terms of the behavior and composition of the targeted fish stocks; the limited extent of capitalization of the fishing fleet; the geographically remote and localized nature of the fishing activity; and the particular social and economic value of the fisheries to smaller coastal communities. As a BIM consultation document put it, the comanaged approach is a "form of *experimental adaptive management* [that] is highly suited to stocks which are structured geographically, and where the relative effects of fishing and environment on catch rates are *unknown*" (Tully 2004: 4; emphasis added). Its management plan explicitly recognizes the importance of managing lobster stocks on a local, territorial level because of the changing characteristics of these lobster stocks and fisheries: "Geographic differences in catch rate, size structure, size at maturity and possibly growth rate and egg production exist in Irish lobster stocks. These differences need to be incorporated into the set of regulations that might be used to manage lobster fisheries" (Tully 2004: 4). It was the incorporation of this knowledge, Carney told me, that determined the geographical scale of the eight territorial units: "we asked: at what scale will the fishermen and the lobsters be enabled to organize and engage in productive relations with each other?" (2009). Carney went on to explain what he meant by "productive relations with each other" by referring once again to the example of the Maine lobster fishery. Marine scientists in Maine

> were using a recruit model as a reference point—that is, comparing the eggs produced by an individual lobster compared to what it would have if it wasn't fished. And the egg production is really low, so the prediction models said the stock was about to collapse. But it didn't because every time they [Maine lobster fishermen] put out more effort, the landings were going up lineally—which is a sign of

> underexploitation, . . . a complete contradiction of the science. I think they reached a decision eventually because the state science and the industry were completely at loggerheads with one another. The scientists couldn't do anything, their recommendations weren't followed through. (Carney 2009)

Carney explained that the problem in Maine was not the data that the scientists were collecting but *the model of science itself*, the comprehension of how the fishery operated and what caused it to fluctuate.[7] Only by abandoning this model could the marine biologists offer a different understanding of the reality of the fisheries, one that could account for its fundamental uncertainty and complexity. This is precisely what happened. Carney told me that the scientists took a step back and consulted the fishermen. The result, Carney concluded, was that "they abandoned the model and suddenly, like a chain being removed, all of these new ideas came forward, new models, new data provision, better indicators, more cooperation *because the fishermen didn't feel threatened anymore*" (2009; emphasis added).

Carney was recounting James Acheson's classic account of the Maine lobster fisheries and their transition to a comanaged model. Acheson was one of the original researchers in Maine who identified the limitations of the recruitment model for describing what was going on there. Despite significant levels of exploitation, data indicated that catch levels in the Maine lobster fishery had been stable since 1947 (Acheson 2003). This became a point of contention in the late 1980s, when scientists found it hard to explain why increases in fishing effort did not deplete fish stocks. With the support of some marine biologists and researchers (including Acheson), the lobster industry argued that this apparent anomaly—high catch rates and high lobster populations—owed partly to how fishermen managed access to the fishery and the fishing activity itself (Acheson 2003). Clearly, there was more going on in terms of fishing practices and the wider ecology that the lobsters and fishermen were a part of. In the Maine lobster fishery, the situation was resolved by abandoning the population model in favor of fishermen-led management tools that incorporated a wider series of social and ecological factors into the management system.

Historically, rules and norms governing the lobster fisheries along the Maine coast have been generated and regulated by the local communities of lobster fishermen. Acheson argues that the tendency of the Maine lobster fishermen to adopt practices that regulate *how* fishing is done, rather than simply regulating how many fish are caught, appears around the world in small-scale fisheries. These practices are not just about allocating resources within the community but also about sustaining the resources on which the community depends. "These rules are unusual in that they are designed to maintain *critical life processes*," Acheson writes (2003, 228; emphasis added). He suggests, "There is nothing that humans can do to ensure steady supplies of fish. All that can be done to conserve fish stocks is to ensure that fish are protected at vulnerable parts of their life cycle. For this reason, protecting breeding fish, breeding grounds, nursery grounds, and migration routes is especially important" (Acheson 2003, 160).[8]

This sense of collective responsibility toward the natural resources on which a community "depends" has now become the target of policy makers seeking to institutionalize community-based resource management, such as those working on the management of the Irish lobster fisheries. In an article about the use of community-based natural resource management in Upland Southeast Asia, Tania Li identifies the extent to which environmental norms about local community action often underlie these comanagement policies. The basic assumption, she suggests, is that people who live close to a resource and whose livelihoods depend directly on it have more interest in sustainably managing it than the state. The aim of comanagement policies is thus to enable these communities to take more control over their resources and the way they are used, with the implicit discourse of empowerment and community participation never far away. The problem with this freedom, as Li argues, is that access to the resources and decision-making processes becomes tied to a particular environmental performance. Communities receive more control over their natural resources on the condition that they assume responsibility for conserving the resources and for maximizing their economic return. Instead of this being about the withdrawal of the state and the narrowly defined goal of bioeconomic sustainability, comanagement strategies can therefore offer governments "an opportunity to rearrange the ways in which rule is accomplished, while also offering communities an opportunity to realign their position within (but not outside) the state system" (Li 2002, 278). Despite their commitment to bottom-up forms of governance, these strategies of devolved resource management can thus involve the extension of environmental rationalities that only recognize certain forms of environmental performance (Bakker 2008; Lockie 1999; Mosse 1997). As geographer Kevin St. Martin succinctly puts it, this "analysis of the commons most often involves empirically discerning the viability and purity of some existent and local community/commons regime in terms of its environmental sustainability, equity, or economic efficiency" (2009, 500).

The Neoliberal Commons

While Irish lobster associations have been making proposals for some form of managed access to the lobster fisheries since the early 1990s, and a management plan reflecting these demands was published in 2008, nothing has yet been implemented. O'Reilly, who has been involved in fisheries management since the late 1990s, was clearly frustrated: "Here we are at the end of ten years and we don't have a lobster management plan. After all those initiations and iterations of a process that you had fishermen involved in, for the right reasons if you like, in terms of sustainability, viability, both economic and biological" (2009).

The situation suddenly changed in 2012 because of Ireland's commitment to European conservation and biodiversity directives. In 2001, Ireland, in concert with other EU member states, agreed to halt the loss of biodiversity. Under the EU Birds and Habitats Directive, a network of nature conservation sites known as Natura 2000 sites were established around Ireland, including many places in coastal areas. These Special Areas of Conservation (SACs) and Special Protection Areas (SPAs) for birds required the Irish government to collect a certain amount of biological baseline data through its conservation agency, the National Parks and Wildlife Service (NPWS). In 2007 the European Court of Justice upheld five complaints against the Irish government for failing to adequately classify or protect several important areas.

The EU threatened heavy fines unless the Irish government complied with the directives. Previously peripheral areas along the coast suddenly became a priority for the state. The legislation required assessments of all licensed activities, including fishing, taking place within these designated areas to make sure they did not adversely affect key biological species living in those habitats. The need to demonstrate that fishing was being carried out sustainably gave a new impetus to the lobster management framework. O'Reilly told me:

> Ultimately, the driver now for any inshore management will be the environmental, like the Habitats Directive, [or] Natura sites. That's interesting isn't it? . . . So all the efforts we've put in for the last ten years, well not the whole of that, that will all fall into place now, at different stages. Maybe not in the same way as we had conceived it in the first place because they have to, not because they have any altruism or whatever, or even sense of moral duty to the resource and the people that are involved in harvesting it. *It's because there is a huge cosh down and it will result in huge fines if they don't, millions and millions.* (O'Reilly 2009)

Despite claims that devolving decision making to local communities is "empowering," this rhetoric comes at a time when not only external economic pressures but also external environmental regulations are placing greater responsibility on fishermen to limit their fishing and demonstrate their compliance. Just as Sam Lee described the new "levers" putting pressure on fishermen to engage in data collection and improved fishing techniques, Maeve O'Reilly noted that lobster fishermen were coming forward for help developing a management plan because of the threat of the European directives. She said they "will now do anything to be involved." Although inshore fishermen do not have a significant impact on biodiversity, the directive would require them to provide proof of this through some form of monitoring and evidence that the fishery was regulated. The demands of the Birds and Habitats Directive thus overlap with the Environmental Management System discussed in the previous chapter.

In the end, European sanctions for noncompliance with Natura 2000 did not materialize, and the plans to manage access to the lobster fishery were put on hold once again. In a subsequent email exchange in 2014, O'Reilly told me that the management plan had now come to a standstill partly because of government "foot-dragging," but mostly because of the "short-term" attitude of lobster fishermen themselves; the lack of support for managed access by lobster fishermen meant that the plan had "run aground," as O'Reilly put it. When I pushed her to specify who these fishermen were or why they were against a system of managed access, she told me they were largely the established, full-time lobster fishermen with polyvalent licenses (access to both nonquota shellfish and quota finfish), who were most dependent on the health of the fish stocks. They were fearful that the new management plan would reduce the value of their tonnage. Tonnage represents the share of the quota allocated to a particular boat: the larger the tonnage the higher proportion of the allocated quota a boat will receive. Tonnage can be sold between boats, but unlike tradable quotas, the transfer of tonnage tends to be a one-timetime event when a fisherman leaves the fishery. In the absence of a pension, fishermen commonly see tonnage as a capital asset that can be realized when they retire from fishing. The reason why polyvalent license-holders in the inshore fisheries opposed any system of managed access to the lobster fishery was that any limits on who can come in and out of the fishing zones would mean a reduced market for their tonnage.

O'Reilly believes that an economically and biologically well-managed fishery should outweigh the asset value of an individual's tonnage. The lack of support from these established inshore fishermen was, for her, a good example of how individual self-interest trumped collective action and the need for short-term sacrifice. She had alluded to this outcome in the first interview I had with her, before the management plan had collapsed:

> The absolute challenge is to deliver sustainability. And I know everyone says it, but it does mean taking strong measures, [and] putting in place appropriate measures and controls so that rules are respected. *That fishermen recognize they do have to be restricted and there isn't mad conflict, people ducking and diving to get around the system.* And that's about communication, and I suppose there is a huge challenge there in delivering it. (O'Reilly 2009; emphasis added)

Consistent with the liberal responses outlined so far in this book, the problem of resource depletion is not connected to the commodification of marine life or the globalization of seafood markets but to the "short-term" (economic) attitude of individual fishermen and the failure to account for it within an appropriate regulatory context. In response to questions I asked about the low prices lobster fishermen received for their catch and the economic anxieties they were likely to have about the future, Brid Smith said,

> So people might say, "We'll manage the fishery, [but] what's that going to do about the price coming in?" Anything that increases your security in terms of going out and catching that fish decreases your costs because it stabilizes the situation, it stabilizes your effort, [and] increases your profitability. So, yes—you want to get a high price for your fish, but more often you want a price that is relative to your outputs, which is making you economic. Which is one side of it, *and of course higher price is something you always want to achieve, but then in a world economic climate that's not necessarily realistic. It's about reducing your costs on the other side.* Currently you have guys who may have up to eight hundred pots. They might shoot half on crab and half on lobster. That still means that 800 pots have to be baited every second day. You're not going to pull 800 pots without a crew-man who has a share. You have to spend on diesel to get you to your pots. You have an engine of horsepower that is not economic in fuel costs. So all of your costs have ramped up. There's no way you can climb down from 800 pots. Then you have to replace gear every year. So your costs have grown. We did some economic stuff. The guys that went over 150 days a year didn't necessarily make more money. (Smith 2009; emphasis added)

This response is similar to the arguments made by Jean O'Sullivan and Simon Casey about the EMS and eco-labels as ways of "arming" fishermen against the globalization of seafood production and the competition that entailed. Instead of attempting to challenge the status quo, fishermen are encouraged to work on what they can influence: the efficiency and profitability of their fishing activity. It is the persistence of this motivation, the need to devise *better* ways of governing individual conduct in relation to the fish stocks that characterizes this approach as a (novel) form of liberal environmental governance; rather than a withdrawal of the state, there is a clear extension of the governmental logic that seeks to fix environmental problems by shaping spheres of activity toward the measurable goal of bioeconomic sustainability. This was made clear by Carney: "If you look at the problem with the lobster fishery at the moment, it's not due to bureaucracy, it's due to a *lack* of bureaucracy, a lack of collective action and talking. If you look at attention paid to it prior to 2000, it was almost zero. I don't think there was a scientific paper published between 1963 and 1995, or a policy paper. *So it's a blank sheet from the state's point of view*" (2009; emphasis added). The recent problematization of the lobster fisheries has challenged this inaction and provoked the need for new and more effective ways of governing the inshore fisheries. Instead of importing existing institutions, fisheries managers are devising new institutional frameworks to reshape the specific social, economic, and environmental conditions that exist in the small-scale, mixed inshore fisheries (Hanna et al. 1996; Crean et al. 2000; Kooiman et al. 2008).[9]

O'Reilly was critical of the Environmental Management System (EMS) because it put an unrealistic burden on the shoulders of individual fishermen. She argued that the comanagement system in the lobster fishery was more sustainable because it could generate necessary forms of "social infrastructure" that would allow designated fisheries to market their catch:

> Another thing you would achieve by having units is an identity and a cohesiveness within a unit which would eventually create a sort of

> *social infrastructure* to kind of carry on to maybe marketing through a certain thing, or branding. And that might generate a shore-based job to do that. But there is absolutely no way of getting to that unless you have a common goal, and those common goals . . . might include the biological, social, and economic sort of thing. (O'Reilly 2009)

The security promised by the exclusive right to fish a particular stock of lobsters enables fishermen the space to *collectively* plan ahead and engage in strong conservation measures. Rather than competition being fostered and harnessed to achieve environmental goals, cooperation becomes the target of governmental practice. As the document "Managing Access to the Irish Lobster Fishery" states, "The objective of managing access is to *enable* lobster fishermen to adopt measures that will give them higher net profits for each pot they put in the water" (Bord Iascaigh Mhara 2008; emphasis added). Restricting access is thus only a necessary starting point, a means to an end. Carney best captures the janus-faced nature of this boundary-making: "It's a system of managing access really, rather than a system of closing off. So instead of having complete open access and anyone getting in whenever they want, *it's managing for the people who are inside in a controlled way, allowing objectives, whether social, economic, or environmental*" (2009; emphasis added). He went on:

> *I think one is trying to stabilize a certain performance.* I think you have to take as the first primary objective the maintenance of economic viability, and still accommodate the social element. *Economic viability to a certain degree in a wide a fishery as possible.* And how do you do that? You increase the catch rate because the catch rate is one fundamental component of the profit equation. How do you increase that? You maintain stocks at a higher level than they currently are. How do you do that? You have to probably gradually get effort back down because all they're doing is competing, the pots are competing for lobster. . . . So how do you bring the pots down? You need to get a collective action. How do you do that? *You create limited entry so the fishermen left have the security to do that, they can make a collective*

> *decision: well now we are a group, we all know one another, we'll agree to a 5 percent reduction per year. So gradually you're reducing costs, you're bringing up the stock, the economics.* (Carney 2009; emphasis added)

Rather than seek "optimal solutions," the managers of institutional commons are looking instead to foster spaces in which communities can take on more responsibility for their own environments and livelihoods while remaining flexible and adaptive. This is the value of collective action: creating "room to develop and tie together promising novelties, thus producing a double capacity to deliver, or as local discourse has it, 'to do it better' than can be done through the unmediated imposition of regulatory schemes" (Van der Ploeg 2008, 202). It is no coincidence that community-managed approaches to resource management have emerged at a time when many national governments are reducing their financial or institutional capacities to respond to proliferating and complex environmental problems. In the context of development policies in the Global South, for example, Forsyth and Johnson write, "For donors and NGOS, [Elinor] Ostrom's design principles offered a model of decentralization and local resource governance that could be replicated in multiple field settings, and which used empowering local and incentive-based governance mechanisms" (Forsyth and Johnson 2014, 1098).[10]

The flipside to this investment in community management, of course, is that the failure of inshore fishermen to act rationally is deemed a failure that demands more work and coordination to overcome. Carney captured this mentality when he told me of the chronic problem of "uneconomic" conduct within the inshore sector:

> *They are not aware of profitability—well, they are aware, but they don't act as if they do.* You could know at any given Monday morning when you start work, they should know what the likelihood of making a profit in the week is—they know the market price, the catch rate from last week, they know their costs. That's a simple kind of profit equation—they can work that out in their head, but when you ask them, "Do they ever work it out in their head?" they never do. (Carney 2009; emphasis added)

Another criticism leveled at the inshore fishermen was that they did not like the idea of being limited within certain boundaries. O'Reilly said that some of the fishermen were reluctant to be "boxed in" within the management units: "All you want to do is say, 'This is the area you mainly fish in, you get to decide how many pots should be fished in that area including yourself.' *It's not about limiting you as much as saying address the issues in this area, concentrate your thing here, what do you want, what are your targets, do you want to increase catch rate and so on*" (2009; emphasis added). What is interesting about these observations is that both Carney and O'Reilly were committed to finding a way around them. While they were frustrated and disappointed about the failure of the management plan (to date), they believed the only answer was to continue working through the various "obstacles" that were arising. How can we get fishermen to be more aware of profitability? How can we get them to concentrate on their local area? In other words, how can we get them to start thinking about their activity (and thus changing their activity) in the bioeconomic terms that are necessary for effective management of lobster stocks?

The literature on the "institutional commons" rightly seeks an alternative to the powerful narrative of the "tragedy of the commons." By paying attention to many successful examples of resource use and management from around the world, scholars began telling a different story. The problem is that the story fails to shake off its liberal origins. When inshore fishermen do not appear to be thinking and acting in terms of economic calculation, they are "lacking awareness," and when they do appear to act "rationally" in so far as they are cautious about the devaluation of their future capital assets, they are identified as being "short-termist." In the familiar language of "improvement," fishermen are framed as individual economic actors who must weigh the potential costs and losses of every action (Mansfield 2004). The result is a call for institutional parameters and norms that can account for and shape this economic conduct within the narrowly defined goals of environmental sustainability (De Angelis and Harvie 2014).[11] The individual fisherman is abstracted from his social and material context and reduced, again, to *homo economicus*. As Agrawal writes, "One of the most neglected aspects of resource use and management in the commons literature is the changing relationship between the environment and human beings who use environmental resources. *If commons scholars consider politics only through the prism of institutions, they fail to attend to human subjectivities in relation to the environment more or less completely*" (Agrawal 2001, 258; emphasis added).

The institutional approach to the commons thus fails to recognize the complexity of social and ecological relations that exist within local contexts—relations tying fishermen to the places they live and work and to the social and natural resources they rely on and co-produce. The reduction of "thick" social and ecological contexts to economic and biological indicators and goals gives rise to new forms of enclosure and "improvement": the apparatuses of liberal governmentality draw and redraw the boundaries of what counts and what does not in the path toward sustainable fisheries. St. Martin writes, "The assumed subject of fisheries (the utility maximising competitive individual) and the space within which that subject operates (the open access commons) have the effect of erasing and/or displacing the cooperative and territorial practices of fishermen embedded within fishing communities" (2007, 543). The erasure of these situated attachments, ecologies, and practices and the different subjectivities and natures they constitute has been central to the modern development of biopolitical and capitalist enclosure. These seem particularly worth examining in a context where the Irish inshore fisheries have managed to remain relatively sustainable into the twenty-first century without much support or regulation from the state or any formal property-rights regime.[12]

5

The More-Than-Human Commons

From Commons to Commoning

A World of Fine Difference

The first time I met a fisherman in Castletownbere was through an incidental exchange in the local pub. I asked the man beside me if he knew any lobster fishermen. He told me to get in touch with another man and gave me his number. I rang him the next morning and met him the following afternoon. It was a Friday and the pub we met in was filling up. A large man sat at the bar and knew who I was immediately. He told me that his wife and sisters were behind us. They met every Friday for lunch. A group of about six women and several children sat over his shoulder. His wife carried a small baby, their grandchild, in her arms. He got me a cup of tea and we talked for an hour or so before he asked if I would come around with him in his van as he had a few jobs to do.

We drove to a small pier about ten miles away. He had some fishing pots to collect. He spoke most of the time. As we drove along the road, the conversation fell on particular points and places of interest: the buoy in the bay that marked where he kept his lobsters, "because Blackball Harbor was sheltered"; the Martello tower built by the English during the Napoleonic Wars above the harbor; the recent weather that prevented his son from fishing on the trawler; the house where his brother lived. At the pier, he asked me to help him pull his small punt out of the water, and we removed the engine together. On the way back to town, we stopped at his friend's house and had tea and biscuits. We sat there talking for about an hour before he dropped me home.

I went out fishing with Tom several times over the next few months. Usually he called me when there was some work to do that involved "a few pairs of hands." A friend who had joined us for tea and biscuits at our first meeting was often there too. As well as going out to sea, these trips would usually involve traveling somewhere to pick up something or other—pots stored on an abandoned pier, salted mackerel kept as bait in a hold by the fish factory, sandwiches from the local shop. I would also see Tom as he cycled on the road or drove past me in town. These different, shared experiences became layers of a growing relationship that was material as much as it was social: eating freshly cooked prawns on deck under a brilliant winter sun after six hours of potting, cutting a dead seal free from a tangled net in the rain, seeing the yellow house he was born in from a mile out at sea, meeting in the supermarket queue and discussing the state of the country, or a chance encounter in the woods when we had not met for a while and we had news to share and he would be reminded I was still around and invite me to go out fishing the next day. Through each new encounter a different story was told, a different reflection or association, about his first home, the boat he used to own, or the places he used to fish. These were all part of the meandering ways in which our relationship evolved and how a place came to be saturated with significance.

What do these everyday, scattered events and encounters have to do with neoliberalism, nature, and the commons? The longer I spent in Castletownbere, the more I grappled with this question. The thickening of my relationship with Tom revealed the rich, lively, and material ways people relate to each other in and through the place they live and work. The problem was that while this ongoing activity was clearly important, it was hard to pin down or even describe. Tom was not just an individual fisherman exploiting fish stocks. He was part of a collectivity that was both human and nonhuman, that did not easily allow for a separation of the social and natural, the material and the immaterial, the past and the future. The ways of knowing and doing that mattered in this world did not translate easily into the terms of political economy or liberal frameworks of governance discussed so far in this book. At the same time,

I don't want to romanticize fishing communities like Castletownbere: fishermen like Tom are not living in some form of noncapitalist, nonliberal enclave, a marginal community that has remained free from the taint of capitalist modernity. Nor are these situated forms of sociality and knowledge entirely ignored by new forms of environmental governance. As I have shown in previous chapters, the relationships fishermen have with their environment and with those around them have become the target of policies aiming to incite and regulate sustainable "stewardship" of the oceans.

Although the modernization of the fisheries has certainly hastened the demise of subsistence fishing, subjecting marine resources and the activities of the fishermen to the logic of the commodity market, this process has not been entirely successful (Palsson 1991; Burkett 1999). The "fisheries are stubbornly resistant to adopting relations of production that are typically associated with capitalism, and *an enormous effort is needed to rein in and discipline fisheries such that they too mirror the singular and hegemonic image of the capitalist economy*" (St. Martin 2000, 960; emphasis added).[1] Vast differences remain between different forms and scales of fishing. The fact that less capitalized forms of fishing continue to exist and provide a living for fishermen reflects the continuing persistence of forms of production that do not rely on ecologically destructive practices. This insight is often missing when overexploited lobster fisheries are identified as needing regulatory institutions to *limit* the economic activity of individual fishermen. What is left out in this analysis is a deeper anthropological account of the complex social and ecological knowledge, practices, and subjectivities that have enabled the inshore fisheries to remain relatively sustainable up until recently.

The missing history of Irish coastal communities and their situated forms of knowledge and value is not new or surprising. John Mac Laughlin writes how fishing communities "clinging to the Western seaboard" were barely visible in colonial and nationalist narratives of development and "modernization." A large part of this was due to their lack of property rights, the marker of citizenship. Like other "nomadic" communities throughout history, vagrants, travelers, and pirates who had no settled "proper"(tied) place with which to identify were hard to identify and control. The German political geographer Friedrich Ratzel argued that the lands of nomadic peoples could be legitimately taken on the grounds that Western, capitalist nations had superior environmental practices. Fishermen seemingly inhabited a "rootless and amphibious world" (Mac Laughlin 2010, 336). As Mac Laughlin concludes, "Because it was accepted that poor fishing and farming communities lived outside modern time and space, they too could be categorised as 'people without history.' This, in turn, hastened the demise of peasant cultivation and subsistence fishing in remote coastal areas, causing impoverished coastal communities in particular to be banished to the footnotes of mainstream history" (2010, 39).

The "invisibility" of peasant and indigenous cultures and knowledge has been well documented by historians and anthropologists (Brody 2002; Rose 2006; De la Cadeña 2010; Escobar 1995; Linebaugh 2008; Merchant 1980; Thompson 1993). The long history of colonialism begins with the erasure of any existing claims to territory or history on the part of those who are being colonized; the concept of *terra nullius* refers precisely to the identification of "waste" land, or land that has not been inscribed with human culture and practice. The term applied not only to the conquest of territories in the "New World" but also to the enclosure of common lands, moors, and heaths that occurred in Britain during the eighteenth century (Goldstein 2013). These processes of expropriation depended on the separation of the spheres of reproduction and production—the "invisible" activities necessary for the ongoing sustenance of bodies, communities, and environments from the "visible" activities necessary for the making of commodities and surplus value.

Theorist Silvia Federici argues that such separation has always been a necessary justification for the violence of enclosure and "improvement." She reads this separation-through-enclosure as something far more fundamental than simply the privatization of land. The relegation of "women's work" (childbirth, child rearing, cleaning, cooking, caring) to the domestic sphere outside the "productive" economic sphere represents the "naturalizing" of this kind of labor. As feminist thinker Maria Mies writes, "All the labour that goes into the production of life, including the labour of giving birth to a child, is not seen as the conscious interaction of a human being *with* nature, that is a truly human activity, but rather as an activity *of* nature, which produces plants and animals unconsciously and has no control over this process" (1998, 45). While the concept of reproduction is most often associated with human reproduction and the management of the "household," reproduction also extends beyond the confines and walls of the house. Federici recognized this after spending time in Nigeria observing and documenting the labor and activity of women in mostly subsistence economies (2012). There, the household, or *oikos,* was not just a home or family but a wider sphere of communal reproduction that involved direct relations with the land, water, plants, and animals.[2] In this analysis, the dynamics of enclosure and capitalist expansion necessarily devalue and erase the myriad relations and practices of (re)production that exist *between* people and the many resources they rely on (De Angelis 2007; Federici 2001; Shiva 2010). By describing and attending to these everyday forms of (re)production, we can begin to pierce the continuing separation of the social and natural spheres and the forms of enclosure this gives rise to. As historian Peter Linebaugh concludes, we need to recover "the suppressed praxis of the commons in its manifold particularities" (2008, 19).[3]

Although critical political economists have analyzed and unpacked the social relations and ecologies of capitalism, there has been less work examining the situated praxis of the commons (Blomley 2008; Bresnihan and Byrne 2015; Mackenzie 2010). As I illustrated in the last chapter, the liberal epistemologies and methodological individualism that characterize institutional approaches to the commons can miss how people relate to one another and their environments in significantly different ways.[4] The historic invisibility of the commons demands a rupture in our epistemological and methodological frameworks: "Before we can reclaim the commons we have to remember how to see it," wrote journalist Jonathan Rowe (2001).[5] The challenge is how to identify and describe these relations, practices, and subjectivities when the conceptual tools we possess are largely derived from a worldview that ignores and devalues them.

Successive phases of enclosure and "improvement," including those described so far in this book, assume the need to shape individual economic behavior by delimiting and constructing spheres of activity that are measurable and thus manageable. In this chapter, I argue that "commoning" necessarily begins from a different epistemological, even ontological, starting point. The commons should not be understood as a limited stock of "common-pool resources" that needs to be protected from individual, human self-interest. Instead, the commons describes all that is co-produced and shared between particular collectives of humans and nonhumans on an ongoing basis. It is not a "thing" but the mesh of humans, animals, plants, land, technologies, and knowledge that enables the making and sharing of "things." As Linebaugh explains, "To speak of the commons as if it were a natural resource is misleading at best and dangerous at worst. *The commons is an activity and, if anything, it expresses relationships in society that are inseparable from relations to nature.* It might be better to keep the word as a verb, rather than as a noun, a substantive" (Linebaugh 2008, 279; emphasis added). "Commoning" denotes the continuous making and remaking of the commons, which starts with the immediate and intimate understanding that the world is *already shared.* As De Angelis writes, "It is in the domain of the shared that limits can be set," and "the dimension of what is shared and of how [it] is shared must go hand in hand" (De Angelis 2007, 241–44).

The perspective opened up through the concept of "commoning" diverges significantly from the liberal frameworks that shape and structure dominant responses to ecological crises. In previous chapters, I have traced how different apparatuses of governance have emerged that both assume and actively construct individualized, economic subjectivities and biophysical "natures" and, through a series of analyses, policies, and practices, harness these to achieve environmental and economic goals. In one sense, "commoning" describes a similar process: as an ongoing activity embodying particular ways of knowing and doing, it also constitutes subjectivities and worlds. But unlike liberal governmentalities, commoning does not operate on or through individualized, human subjects plotting their actions on the basis of economic reasoning. Commoning, in my interpretation, begins with a subjectivity that is already in the thick of things, already mired in a social and material world that is relational and, crucially, shared.

There are fields of research that can help us better understand how commoning departs from liberal accounts of experience, subjectivity, and nature. These include the work of anthropologists, who examine indigenous cosmologies and relations with nature and territory (Escobar 1999; Viveiros de Castro 1998; Rose 2004), and more-than-human, feminist, and new materialist theory that shift the methodological and epistemological lens away from discrete, calculating human subjects and objects to the *relata,* the relations that constitute our world in fundamentally material ways (Barad 2004; Bennett 2010; Puig De la Bellacasa 2011, 2012; Haraway 1988; Papadopolous 2010a, 2010b).[6] These literatures enrich our understandings of nonliberal subjectivities and agential "natures" and help us disrupt the liberal humanist epistemologies that both individualize and place humans at the center of world-making processes. This relational perspective shows that knowledge and meaning-making are not the preserve of the individual human actor (and the target of neo/liberal rationalities of government). Instead, knowledge and action are always a collaborative endeavor and develop within certain material limits. These limits are not "natural" limits that are given in advance but material restraints that inhere in any situation and need to be negotiated.

As indicated by the term "more-than-human," the interdependence that ties the commons together is not just a human matter. We cannot begin describing and organizing the world by separating the individual rational subject from nonhuman nature. Even if CPR approaches, for example, attempt to account for the changing, unpredictable dynamics between particular communities of fishermen and the fisheries they exploit, they continue to identify particular "stocks" of fish and the economic fishermen who exploit them. This bioeconomic approach fails to account for the widely distributed *mesh* of relations that exist among fishermen, lobsters, and the wider territory of the commons. Take the simple journey I made with Tom, for example: from meeting in the pub with his family, to the sheltered harbor where he kept his lobster catch safe, to the help he required taking his boat out of the water, to the casual conversations we had with his friend over tea, a friend who came out to work on the boat a couple of weeks later. Tom is both a part of, and reliant on, these different "resources" (family, harbor, boat, fish, friends), resources that are vital to everyday social and economic life. The messy work involved in negotiating and maintaining these relationships cannot be explained by, or reduced to, the economic reasoning and practices of liberal governmentality.

In the first part of this chapter, I want to convey some sense of the unpredictable agency and vibrancy of the marine environment in which fishermen operate. The intention is not to emphasize the uniqueness of fishing as an activity but to indicate the ways in which "nature" is not experienced as a fixed or even knowable domain separate from interaction with it. Although new modes of governance have recognized the uncertainty and unpredictability of the sea and the fisheries, the goal underlying such "adaptive" management remains one of control: to render it amenable to regulation and profit-making. These perspectives continue to operate on the basis of a dualistic understanding of society-nature that situates knowledge of "nature" in the heads and bodies of individual human actors, not in the unfolding relations between humans and nonhumans. The fishermen I met and worked with, however, accepted the constant, material uncertainty of the marine environment. Their negotiations with these everyday uncertainties—the sea, the elements, the fish, and technologies—reveal a more distributed and materialist concept of agency. This relationship is not stable or consistent but relational, in the sense that it changes from one context to another, drawing out different responses and effects that cannot be known in advance. This contingency amounts to a very different knowledge of, and relationship to, the nonhuman that resonates with indigenous conceptions of territory and nature as vibrant, living environments that include humans.

The concept of the more-than-human commons indicates the *interdependence* of humans and nonhumans. In the second part of the chapter, I examine how this recognition requires inshore fishermen to take care of the fisheries they rely on. This "care-work" can only be explained in terms of an ongoing, situated activity rather than something that can be institutionalized through individualized access rights or auditing technologies like the EMS. Feminist conceptions of "care-work" allow us to see that the activity of caring is always situated and relational and resists valuation through techno-cultures that attempt to quantify and measure it. More fundamentally, the concept of care, which is central to the activity of commoning, asks us what is needed to account for and hold together multiple needs and agencies without submitting to the desire for control. It reclaims a type of agency that depends on our interaction with others, rather than omitting these interactions through a false, anthropocentric fantasy of resolution.

The social and material importance of practices of mutualism and reciprocity in subsistence economies has been well documented. In the third part of this chapter, I consider their function within Castletownbere. Countless acts of gift-giving, favors, and loans made up an extensive and almost invisible network of exchange that, very simply, allowed things to happen. Significantly, the social nature of these forms of exchange disrupts the binary of altruism or utilitarianism that tends to explain (away) the motivations for different actions by resorting to the figure of the liberal decision maker. The production and circulation of surplus through commoning helps us move away from standard liberal economic explanations that orientate around the management of scarcity. From this perspective, the circulation of surplus through the commons is both a contribution and investment in the commons; doing some work for free or lending somebody a tool is neither an act of charity nor a calculated economic exchange. As with the care-work described in part 2, the rationale for gift economies in a place like Castletownbere is largely a pragmatic and material response to specific and often unforeseen needs. In the final section, I elaborate on this pragmatic mutualism in terms of the "invisibility" of the commons and the challenge of articulating it outside the norms and values of biopolitical control and capitalist enclosure.

The More-than-Human Commons

The first time I went out on a trawler, it was early September and the tuna were migrating up the west coast of Europe. Irish boats were going out again for tuna after a multiyear ban on tuna fishing, and fishermen were hoping to reap high profits. My neighbor knew the skipper of a trawler from Rossaveal in Galway that was going out from Castletownbere, and they were short a berth. I was told I could go out if I felt up to it and was ready to work.

My neighbor called on a Wednesday to tell me the trip was delayed because someone the crew knew had drowned in a boating accident, and they had to go back to the funeral in Galway. I was told to go to the quays on Friday afternoon to head out that evening. The weather had been good up until that point, and quite a few of the boats in the port were out, some of them after the tuna. We were going pair-trawling, which meant two boats went out together and pulled a net between them to increase the catch. I met the crew of both boats as we waited on the quay to hear whether we would be going out. The crew consisted of the skipper's family from Galway, locals who were out of work, and a couple of young Polish men. We waited for four hours as the crew kept in touch with boats that were already three hundred miles out to sea at the fishing grounds. They were talking to different boats to see whether they should head south to the Bay of Biscay, where they were recently, or out west. I asked Conor, the son-in-law of one of the skippers, whether he knew how long we would be out. He said it was impossible to know. They had been out ten days before and caught nothing. Huge waves had covered the boat and washed away the lifeboat. He had cancelled his family holiday to Portugal though, because he could not be sure the fish would be back. "It all depends on the fish," he said.

At eight in the evening, Conor informed me they would not be going out because the tuna were scarce. One of the Polish men cursed because they had been tied up for ten days in Castletownbere already and had no money because the fishing had been so bad. Conor took my number and said he would call tomorrow if they got any good news. Because tuna can only be caught at night, the crew would either have to wait until tomorrow evening or just return to Rossaveal.

The next day, I heard nothing so I drove by the quays to see if the crew was still there. One of the crew was looking over the net. He told me the reports had not been good. On Sunday morning, I got a call during breakfast. Conditions had improved: the water was warmer, and a boat had caught twenty tons of tuna the night before. I grabbed my bag, which was still packed, and met them at the quays. In beautiful weather, we left the port with the other boats. Everyone was going fishing.

At the time, the tuna were three hundred miles southwest of Ireland. It took us three days to get to the fishing grounds. For two and a half days, we did not see a single other boat. The boredom I had heard about was palpable: all we saw besides the sunrise, the sunset, and an expanse of sea was a shoal of over a hundred dolphins that bounced toward us across the flat sea one morning. They dived around the prow of the boat before setting off in the opposite direction.

We arrived at the fishing grounds in the afternoon, so we had to wait until it got dark before putting out the nets. Everyone except for the skipper got some sleep. At 11 p.m. I awoke to the cry of "battle-stations." The scene that greeted me on deck was entirely unexpected: there were about sixty or seventy lights of other boats scattered in the black night sky. It was hard to distinguish where the sky ended and the sea began. The effect made it seem as if we were in space.

The crew started the winch to let out the nets, but it did not work. There was much cursing and tinkering with the machinery, but they did not know what the problem was. They had checked everything on shore before coming out, and there had been no problem. The skipper said: "If it's not the fish it's something else." It looked like we would have to go back in—the second trip without landing a fish. I was told to go back to my bunk to get some rest but was woken an hour later: they had called a fitter on another boat who suggested the problem might be an air leak. They released some oil and sure enough it worked. By then it was about 2 a.m., and there was still enough time to get the nets out.

Because we were pair trawling, we had to approach another boat that had come with us from Castletownbere and attach ropes and a net. These are the notes I wrote the next day:

> We came up close to the other boat. It is hard to describe how cinematic it was: the boat lit by lights in the midst of the black sea, waves and wind whipping around it, three men in yellow oil skins on the back deck. The boat was only fifteen feet away or so. It seemed very dangerous—controlling the boats so well in the conditions. As soon as we were close, something fired across—I didn't see it—and thumped down beside me. It took me a while to fetch it, by which time Jimmy had pulled it in anyway. I pulled away at the slack of the rope as Conor continued to pull it in. Everything was frantic. There was much shouting and speed—to get the net attached as quickly as possible because of the danger of the boats being too close to one another. Conor shouts at me to help Jimmy but I can't hear the instruction in the wind and the noise. The fourth time I hear and help him but by then we were already losing it (I couldn't see exactly what was happening but the rope started firing out the back of the boat). The rope slipped through my hands and cut my fingers. Jimmy pushes me out of the way because I am standing on a coil of rope and he quickly says that's "deadly dangerous—always watch where your feet are, you'd be whipped out and dead before we got you back in." The cable is lost and Conor curses. The boat moves forward and then stops and the stern of our boat is only a foot or so from the front of the other one—Conor is screaming at Paddy to go forward as the boats look set to collide. The channel between the boats has caused the waves to funnel up so they are nearly coming in over the deck. Paddy can't go forward because the throttle cable has snapped. Eventually the boats part. Conor went down to fix it before we start again. (Bresnihan 2009a)

Conor finally fixed the throttle cable, and we managed to get the nets out with the other boat. Once the net was out, we just had to wait for the fish. Each trawl takes about four to five hours. It was impossible to know what would be in the nets. The previous night a boat that had been fishing the same waters had caught thirty tonnes. When we pulled in the nets at 6 a.m., there was not a single tuna. After all the trouble with the equipment, it had been too late for fishing.

The following two nights we caught a few tonnes of tuna, but the other boat pulled them in. In order to balance the loads, we took the catch on the fourth night. It was the first time I had seen nets being pulled in out of the sea. When the nets slowly came in, it was dark and there was just a glimmer of a sunrise on the horizon. The bulk of the fish collected at the net's "cod-end," which is hoisted over the hold before being released. As soon as the net opened, a load of fish came slithering out. In that single catch, we landed about three hundred fish. Most of the fish were battered, scarred, ripped, and even unrecognizable. Some flipped about, while others had guts hanging out. We waded through them up to our knees for the next three hours as we sorted and stacked them.

The skipper of the tuna trawler, Paddy, was in his seventies. He was originally from the Aran Islands off the coast of Galway. He had started fishing in a *currach* when he was twelve.[7] Even as the skipper of a fifty-foot trawler with sonar technologies, he still confronted the unpredictability of the sea, of fishing—something he admitted to me himself. I had imagined that fishing on such a large boat for tuna would be easier than fishing on the smaller, inshore boats that I had become more familiar with. I had presumed that sonar technology would allow fish to be targeted with ease. Although the scale of it was certainly different, the basic relationship with the elements, and the fish—the uncertainty—was much the same.[8]

During my fieldwork, I met social and natural scientists who were fond of comparing fishermen to hunters on the basis of their "intimate" knowledge of the seas. Scholars have explained the need for fishermen to be attentive to the changing environment as a form of situated or embodied knowledge, a way of knowing through doing (see Palsson 1993). This emphasis on "practical" learning, or apprenticeship, suggests skill emerges through a bodily engagement with the world that extends beyond language or symbolic representation.

Anthropologist Timothy Ingold describes this process as a "sentient ecology"—not just a formal, authorized knowledge of the environment, but a "felt" awareness developed through long experience in a particular environment (1986, 2000). He notes, for example, how a biologist regards the tree as an inanimate object, whereas the hunter, "accustomed to the woods," registers the tree through "the swaying of the boughs in the wind, the audible fluttering of leaves, the orientation of branches to the sun" (Ingold 2000, 98).

This emphasis on the situated knowledge of the human subject suggests that knowledge develops over time and that a particular way of acting in, and through, the world becomes refined through experience. Such phenomenological perspectives, however, continue to emphasize the human subject as the site of experience and knowledge (Rodaway 1994; Grasseni 2004): the "hunter" analogy suggests that the hunter perceives the tree in a certain way because he is a hunter.[9] By interacting with, and gathering knowledge about, the outside world, individuals cultivate an "education of attention" that allows them to operate effectively within a changing world (Gibson 1950).

The problem with this "shallow empiricism," as social psychologist Paul Stenner calls it, is that it does not go far enough to undo the subject/object dualism that dominates liberal epistemologies. Although knowledge of the world may not be *a priori*, it is still understood to be the possession of an individual human subject who forms an "interior" knowledge of an objective "exterior" world of things.[10] One of the consequences of this dualism is that the "local" knowledge of fishermen can be included alongside other forms of "scientific" knowledge within fisheries management (Berkes et al. 2000). The role of the fisherman as an active producer of knowledge alongside scientists and fisheries managers has been one of the defining features of the transformation in fisheries management over recent years. Fishermen are being encouraged to record and share their experiences and observations through new science-industry forums, auditing technologies, and community-based management institutions. Although these all produce different forms of knowledge about the fisheries and the activity of fishermen, they all share in the belief that the world, particularly the complex, nonhuman marine environment, is amenable to human understanding and ultimately control. When complexity and uncertainty are acknowledged as essential characteristics of the environment, new tools, methods, and resources are mobilized to account for and manage this complexity and uncertainty . Although the fishermen I spoke to also understood that the environment in which they worked and lived was unpredictable, they had other ways of expressing and relating to it that can help us pull away from liberal, humanist understandings of nature, agency, and knowledge.

While scientists awkwardly described fishermen as having "intimate" knowledge of the sea, another more pejorative description was that they were good storytellers. The first time I met Jane Downing, a scientist who worked for BIM and lived and worked in Castletownbere, she asked me how I was going to "truth" the fishermen and suggested that fishermen never told the same story. The idea of fishermen as storytellers is a familiar one, but the idea that the story changes each time it is told reveals something about how "things" depend on their context (Nightingale 2013).

Although it is certainly true that fishermen respond to different signs within their environment and therefore develop knowledge and understanding through their ongoing experience, what struck me most about going out on the boats was the extent to which they *did not* seem to know what was going on from one moment to the next. When I met Paddy, the seventy-year-old skipper, he did not stress his decades of experience but instead the tension of always having to respond as best he could to the various situations that arose. The actions that he took were not really his own nor even those of his crew but responses to, and coordination with, the different rhythms and agencies of animals (fish), artifacts (machinery), elements (wind, tides) and, perhaps finally, people. I lost count of how many times fishermen would shrug their shoulders and say "I don't know" or "that's just the way it is" when asked why something happened or why it did not.

On one of my last days in Castletownbere, Tom, the old, semiretired fisherman, asked me to come out fishing for prawns so I could get "one last feed" of them before going back to Dublin. We hauled four strings of pots we had shot the day before. There were forty pots on each string. We caught nothing except a few small crabs and three prawns. I asked whether it was the weather and Tom said the prawns usually liked an easterly wind. I asked if the pots were not out long enough, and he said they often check them after a day and there are plenty of prawns. Tom had a reputation as being one of the most experienced fishermen in Castletownbere. He laughed and said, "I don't know who to ask about this, three bloody prawns, it must be the worst ever." What I gleaned from this incident and Tom's response was that fishing always involves frustration and disappointment, that the sea is always changing, and that we, humans, are not at the center of that change.

A diverse body of work has sought to "deepen" our understanding of experience, knowledge, and nature by doing away with an analytic focus on the human subject as the producer of all meaning and agency (Barad 2004; Bennett 2010; Coole and Frost 2010; Latour 1993; Stenner 2008; Stengers 2008; Papadopoulos and Stephenson 2006).[11] At the heart of this shift is the decentering of the human subject and an opening up toward nonhuman or more-than-human agency in the making of social life. Importantly, this expansion of agency beyond the human subject does not simply mean that technological artifacts and nonhuman organisms "do things." It instead situates their (and our own) capacity to do things as an effect of relations with other artifacts, animals, plants, and people. This perspective requires the "complexification" of existing relations and contexts in order to account for "the complex associations of entangled, socionatural beings, instruments, and practices that constitute different natures" (O'Reilly 2005, 116). We are always situated in relation to other people, places, and things, and knowledge and agency is constituted through these relations. In this understanding, the primary ontological category is not the "individual" but the "relationship" or "alliance" (Stenner 2008).

Theorist Karen Barad has developed the concept of "agential realism" to capture how knowledge is produced *between* human and nonhuman entities (Barad 1999). Setting out from a critique of liberal epistemology, she argues that agency is always the outcome of a collective "intra-action" between entities rather than an inherent or preexisting property. Her neologism *intra-action* pushes us away from *interaction*, which presumes the prior existence of independent entities. It is in the process of "intra-acting" that subjects and objects come into being. She writes, "Agency is a matter of intra-acting; it is an enactment, not something that someone or something has. Agency cannot be designated as an attribute of 'subjects' or 'objects' (as they do not preexist as such). *Agency is a matter of making iterative changes to particular practices through the dynamics of intra-activity and the mechanism of unfolding*" (Barad 2004, 112; emphasis added).

Perhaps the most interesting work in this area has emerged from the rich anthropological accounts of alternative nature-social relations, specifically research that has examined indigenous cosmologies and notions of territory (De la Cadeña 2010; Escobar 2008; Rose 1996; Viveiros de Castro 1998).[12] Marisol De la Cadeña describes these indigenous relations as the "nonrepresentational, affective interactions with other-than-humans" (De la Cadeña 2010, 346).[13] Philosopher Deborah Bird Rose has written about the nature/culture dualism through her experience of working with Aboriginal people in Australia and the land they consider themselves to be a part of. "In indigenous culture," she writes, "there is no nature/culture divide. One could say that country is all culture, *but the more interesting point is that it is all sentient, communicative, relational, and inter-active. In this sense, culture is not something you have, but rather is the way you live, and by implication, the way your knowledge arises and is worked with*" (Rose 2013, 100; emphasis added). What becomes clear in Rose's accounts is not simply that Aboriginal culture takes seriously the sentience and meaning-making of nonhuman life (through spiritual and material ritual and practice), but that this understanding arises from the immediate assumption that their country is shared with human and nonhuman others. This basic premise does not originate from an idealized ecological "ethic" but from a thoroughly materialist, situated knowledge of their ecology and the complex interdependencies that exist therein. She writes,

"There is nothing 'natural' about the continuity of life on earth, nor is continuity a process which can be taken for granted" (Rose 1996, 44). The Aboriginal relationship to their country therefore expresses a different *investment* in the production of life, a different form of biopolitics that is practiced and organized around an intimate awareness of more-than-human life processes. Rose's term "nourishing terrain" conveys this mutualism by at once referring to the nourishment provided by the terrain for the people who live there, and the nourishment provided by the people for the terrain they rely on. These relations of interdependency should not, however, be confused with a pre-Lapsarian harmony of human and nonhuman. Instead, they reflect the need for different knowledge practices that are capable of paying attention to the multiplicity of overlapping and contested needs and forces that constitute any territory and that need to be *negotiated*.

Interestingly, Rose extends her analysis beyond the Aboriginal case.[14] Referring to the "intersubjective, countermodern, embodied, and dangerous collaboration" of horse and man in rodeo performances, a quintessentially white-settler site of nature-culture, she writes, "'As we hold our breath and clench our hands, we can find ourselves increasingly excited at the thought that maybe, perhaps, civilization will not win, ever" (Rose 2004, 94). She is referring here to the way in which neither the man nor the horse involved in the rodeo performance are "winning" or dominating each other. The aim is to hold on to each other, to come to some sort of understanding that is necessarily ephemeral and difficult. She suggests we should avoid an impossible and idealized binary between anthropocentric mastery of nature or "natural" anarchy (of which *terra nullius* is one obvious articulation). The interaction between the horse and its rider is a performative collaboration that involves practices of attention, a kind of dance that is not designed to end with a single victor. Although this collaboration is never harmonious or painless, it avoids the violence and destruction inherent in projects to "civilize" and "tame" the wild.[15]

In a different context, sociologists Dimitris Papadopoulos and Niamh Stephenson describe this messy, situated relationship with others as a way of "tarrying with time," a break with liberal, linear time and the subjectivity that travels along it: "Tarrying with time does not entail a concrete vision of an alternate future, but an expanded, slowed-down present which fuels new imaginary relations with other actants and new forms of action, possibilities people are compelled to explore, but which only later and unexpectedly will materialise in an alternative future" (2006, 158–59; emphasis added). The future is always unpredictable because it is the continuous result of a shared, collective process of negotiation in the present. Entanglements of objects, animals, and people reoccupy understandings of subjectivity: "Other people, things, material spaces, situations—all these actants—participate in the unfolding of experience. Experience is not primarily a matter of thought. Things and spaces are carriers of experience, which becomes ours" (Papadopoulos and Stephenson 2006, 442).[16] The concept of "tarrying with time" always returns us to the present, to the material possibilities *and* limits that exist here and now; social and environmental change is not just a human affair but something that is always conducted with human and nonhuman others, whether we like it or not.

I interviewed Johnny, a fisherman, one evening in his house. His father was a fisherman. He told me how he had begun fishing as a young boy during the summer holidays, helping his dad catch salmon in July, shrimp in August, and then pollack and crab just after he had gone back to school. He eventually started making his own pots and fishing them himself. He found it patronizing and offensive when outsiders accused fishermen of not thinking long-term. He said fishermen were the only ones who truly had to think about the future because they relied on the fish. Civil servants, he argued, were never in their positions for more than a few years. Why would they care about the fisheries if "they'd still get their paycheck?" At one point he paused and pointed out the window at a black boat that was sitting on a trailer. He had just bought it for his eldest son, who was thirteen.

For Johnny and other fishermen I met, the present is not separated from the past and future in linear terms. It instead gathers the past and future into itself, "like refractions in a crystal ball" (Ingold

2000, 205).[17] Actions in the present tie not just to a future-becoming but also to the past-becoming, a living connection with what came before. All these elements of naming and narrating are bound up with a strong sense of connection that people in Castletownbere felt to the things and places that populated their everyday worlds.

Deborah Bird Rose describes something similar when she writes how a white settler regards his relationship to the place he lives and works as "spiritual." He understands the place, Nature, as an "active force, something living, something to be encountered," even though he is also involved in economic practices that have undermined and destroyed it (Rose 2004, 207). Although fishermen are the primary exploiters of the fisheries, their relations to the marine environment are, to different degrees, more complex than this narrowly economic one. A trip out on a fishing boat reveals a world that is "buzzing with multitudes of sentient beings" with whom fishermen interact on a regular basis (Rose 2013, 95). When a fisherman tells me that "fish are devious," or when a small lobster is thrown back in the sea and told it is getting "another round," or when a seal appears beside the boat and is cursed for "stealing" fish from the nets, fishermen confer an agency on nonhuman life that is not easily representable within the economic reasoning of environmental governance. The marine environment is not just a dumb, external world of resources waiting for intensive exploitation or regulation through highly technical performance metrics. Although fishermen are inserted within circuits of production and accumulation, they are also part of everyday natures that have escaped the enclosure and separation of modern space and time. These everyday natures unfurl through ongoing interaction and negotiation between multiple agencies. In the next section, I will examine how this immediate and intimate sense of a shared world translates into situated practices of care and attention.

Care of the Commons

Although some fisheries managers regard the reluctance of fishermen to support the proposed community-based lobster management plan as evidence of the "short-sightedness" of fishermen when it comes to conserving fish stocks, a different perspective emerges when you talk to a fisherman like Joe, a young, full-time lobster fisherman. While he understood the need for regulation, he was wary of the proposed authorization system because it failed to distinguish between different kinds of fishermen. Granting an authorization to everyone who currently fished for lobsters appeared to draw an arbitrary boundary in the present, excluding considerations of the past and the future that were beyond the economic reasoning of the fisheries managers. These concerns can help us understand what is at stake in the more-than-human commons and what is meant by commoning. Understood as an *ongoing activity*, a person's claim to be included in the commons becomes a question of how they *relate* to, and *participate* in, the making of the commons rather than a formal right vested in the individual.[18]

Joe had been involved in the initial consultations on the management of the inshore fisheries, was secretary for the Irish Fish Producers' Organisation (IFPO), and was one of the national representatives for the inshore sector. He was sorry that the inshore fisheries were in such trouble after ten years of effort to institute a form of comanagement. Like all the full-time inshore fishermen I spoke to, he identified the lack of regulation as a principal obstacle to sustainable fisheries. For full-time fishermen like him, the fact that anyone with a boat could effectively fish for lobster in the summer was a real problem. He was referring to a decision by the Irish government in the early 2000s to grant all nonlicensed fishing boats around the coast a free potting license. The problem was that these part-time fishermen did not rely on the fishery. They were opportunistically catching fish without having to think about the future:

> They could all now fish for free, and many of them had other jobs, fully pensionable jobs as teachers, *gardaí* [police], civil servants, and fished one hundred or so pots in the summer for beer money or whatever. They didn't have many overheads, most of them were in small punts and could handle ten pots on a string, ten strings, one string an evening. *They don't care about the stock.* It is them who hammer the stock because it doesn't matter what happened the next year, they

> don't rely on it. People have been known to leave their pots there [out on the rocks] for a month or so, too lazy to collect them, or hoping to just catch a load in one go. (Joe 2009)

Joe was describing the different *relations* that exist between fishermen and marine resources, relations that are the basis of different social practices and forms of knowledge (De Angelis 2007). In drawing attention to the existence of these different relations we can begin to see how questions concerning the regulation of access to and use of the fisheries are not simply about limiting economic activity or regulating a particular stock of fish. For Joe these questions circulated around conceptions of fairness that had to be worked out in respect to the social and material limits of the place he lived and worked, something that is stripped from dominant forms of bioeconomic resource management.

Significantly, Joe defines the part-time fishermen more by what they do not do than what they do. Fishermen who are not as reliant on the fishery do not need to respect the same limits or follow the same practices as those who are. Joe and other lobster fishermen I met and fished with were acutely aware of the need to look after the lobster stocks, to ensure that they come back year after year. This kind of attention means not only limiting the number of lobsters you take but also checking the types of lobster you take, how often you leave your pots down, for how long, and how many you use. It means protecting selected mature, female lobsters by "V-notching" their tail fan. Holding a fertile female lobster up to me, Tom, a long-time fisherman, said: "A fisherman killing a berried hen is like a farmer killing his cow when she is in calf." The ongoing reproduction of the lobster is something that lobster fishermen who rely on the lobster fisheries for their livelihoods cannot ignore. This is not because they are more "enlightened" and "conservationist," but because they recognize material limits and need to operate within them.

At the same time, fishermen are also opportunists. They operate within limits, but they also seize what they can, when they can. On most inshore boats, fishermen use plotters to mark where they have deposited pots and trawled nets. If one site is "fishing well," then you keep returning to it. Although this technology makes the task easier, it is no different to how fishermen have traditionally worked.[19] One day I was out potting with an inshore fisherman, Brian, and noticed a blue box on his plotter map. He explained that the box was part of a project carried out by BIM to show fishermen that intensive fishing in one area led to declining profitability. They had dropped six strings of pots close together in the area. Brian laughed and told me the study had backfired because the pots kept coming back full of fish. Like all of the fishermen I met, he held to the basic rule that "you fish when the fishing is good." When fish were abundant, you took as many as you could because fish themselves were unpredictable. Time and again I heard and witnessed how fish appeared and disappeared without much explanation or need of explanation. The everyday ways fishermen negotiate both material limits, such as the reproductive cycle of lobster, and opportunities, such as the momentary presence of fish, reveal something that is not captured by simple biological or economic indicators—a relationship to the shifting possibilities that inhere in any particular context.

The more-than-human commons departs from a liberal focus on *homo economicus* in its immediate and intimate understanding that the (localized) world where life and work take place is *shared*. This view contrasts with the institutional approach to the commons discussed in the last chapter, which begins from the opposite assumption: resources are not shared, individuals operate as individuals, and self-interest has to be remedied through appropriate incentives and institutional arrangements. By assuming that the world is already shared, I am not suggesting that individual economic self-interest and the degradation of fish stocks are illusory. Rather, I am proposing that commoning only arises in a context where resources and capacities are already shared. This is not a naïve aspiration or ideal to aim for, but something that is socially and materially constituted in contexts where humans and nonhumans are dependent on each other.

Although this conceptualization of commoning challenges formal institutions of individual ownership, it does not mean that there are no limits on what an individual can and cannot do. The social and material interdependency of the commons *requires* forms of sharing. Rather than shaping individual conduct through liberal practices and institutions, practices of commoning are based on the ongoing recognition and negotiation of limits within the domain of the shared (De Angelis 2007). These limits are not abstract, quantitative figures (like quotas) but concrete material limits that inhere in any ecological collective that an individual is part of and relies on. Because these limits (and opportunities) change, often from moment to moment, the ways they are identified, negotiated, and responded to must be dynamic, flexible, and open to contestation.

Joe, for example, wanted to know how a system of individual authorizations would cater to fishermen who had been fishing elsewhere all their life and now just wanted to retire and "fish a few pots." Such fishermen would not have any track record of being a lobster fisherman and would not be able to afford a license if one became available. But the return and "semiretirement" of older fishermen to the inshore fishery was a regular occurrence in Castletownbere and an important part of the social economy, especially when many of the older fishermen did not have pensions. A liberal framework of individual rights is not able to deal with the social, situated character of this concern.

Fishermen like Joe are aware that they share the fishery with other fishermen, fishermen they know and see every day. One day, a small punt appeared some distance away, obviously fishing pots in close to the cliffs. I asked Joe if the boat belonged to one of the part-time fishermen he was complaining about. Joe immediately knew who it was—a semiretired fisherman who according to Joe knew how to take care of the fish. He knew this because he lived and worked beside this fisherman. This is not mysterious; it is the kind of knowledge that one gathers from everday social interaction and observation. This kind of sociability means that formal rights of access, such as authorizations, have not been required until recently. Instead, informal norms and practices exist, guiding when and where fishermen can drop their pots and nets.[20] In Kerry, for example, where there are even fewer lobster fishermen than Castletownbere, the only practice they observe is a first-come, first-drop rule. In other parts of West Cork where the fishing is busier, however, more stringent norms are required. Twice I encountered cases where a fisherman's strings had been cut by another fisherman. This makes the retrieval of the pots on the sea floor very difficult. It is a practice that local fishermen use to discourage individuals from fishing new grounds without some kind of prior knowledge or discussion with whoever is already fishing there. It is also a practice that is understood and accepted: even though the fishermen whose strings were cut stood to lose several thousand euros, they both agreed it was their fault, that they should not have dropped their pots there in the first place. The almost taken-for-granted existence of these social norms explains, at least to some degree, the wariness that some fishermen feel when it comes to introducing formal authorizations and licenses for individual fishermen, something that does not appear to take into account the practical commitments involved in sustaining the fishery and the way these commitments change over time.

This situated understanding of use and access in the fisheries is contingent on practical commitments to the production and care of the commons, understood not just as a discrete resource but as a wider collective of people, resources, and place. This meaning can already be found in the roots of the word "commons": "com-" (together) and "-munis" (under obligation). First, this derivation tells us that the commons is produced together, reflecting our interdependence and the assumption that our world is already shared. Second, and arising from this, it infers the obligation that such interdependence demands of us. The commons is not a "thing" that we can access through title or authorization but something that is ours because we produce and care for it, *because we common*.[21] This is a subtle but important distinction from the institutional commons approach, but it prevents us from reducing the meaning and value of commoning to a legal or institutional framework—that is, all fishermen are counted as the same and granted a legal right of access to the "common pool resource" by a central authority.

The concept of "care" is helpful in trying to articulate the situated, material ethos underlying the activity of commoning. Care denotes the immediate interdependency of human and nonhuman life; it expresses relations between particular communities of humans and nonhumans that are not fixed or prescribed in advance but worked out according to and across a variety of different needs and interests. Feminist theorist Joan Tronto defines "care" as "everything that we do to maintain, continue and repair 'our world' so that we can live in it as well as possible. That world includes our bodies, ourselves, and our environment, all that we seek to interweave in a complex, life sustaining web" (quoted in Tronto 1993, 103). The idea that care holds the world together by connecting all aspects of our lives returns us to the interconnectedness and interdependency of human and nonhuman life. Such an understanding carries us beyond liberal notions of ethics that focus on the individual and "self-care" to the individual as part of a human and nonhuman collective that must be nourished on an ongoing basis. It also regrounds or rematerializes relations of care; as Rose writes, "ethics are situated in bodies and in time and in place" (Rose 2004, 8). Instead of simply measuring the output of a given activity, care therefore points to the wide scope of attention required to ensure an activity is done well.

Theorist Maria Puig de la Bellacasa uses the example of permaculture to help communicate the fundamental differences between an ethos of care and more normative conceptions of ethics. The ethos of permaculture first of all attracts our attention to the "invisible but indispensable labours and experiences of earth's beings and resources" (Puig de la Bellacasa 2010, 165). Recognition of our reliance on these labors immediately decenters any sense we might have of our own agency and also places us into a relationship with nature that is not abstract but always material and situated. It requires ongoing action and care, not as additional components or prescriptions for our "normal" activity, but as integral parts of how we do things. Such accounts can appear to correspond with neoliberal discourses of environmental stewardship and responsibility. While auditing technologies like the Environmental Management System are designed to document and assess the environmental "performance" of fishermen, however, the attempt to codify and represent the affective, situated, and relational nature of care-work more accurately reveals the gap between attempts to capture and enclose such activity and the activity itself.

Caring does not begin and end with the individual. It involves ways of knowing and doing that always need to be negotiated with others. As Justo Oxa, an indigenous schoolteacher from Peru, writes, "Respect and care are a fundamental part of life in the Andes; they are not a concept or an explanation. *To care and be respectful means to want to be nurtured and nurture others, and this implies not only humans but all world beings*" (quoted in De la Cardeña 2010, 354; emphasis added).[22] In this way, care practices are grounded in a respect of limits. At the same time, however, such limits are never fixed because they are part of a wider ecology of forces, needs, and possibilities. Navigating these changeable parameters is the practical, situated "doing" of care-work. Mabel McKay, a Powo healer, observes: "When people don't use the plants, they get scarce. You must use them so they will come up again. All plants are like that. If they're not gathered from, or talked to and cared about, they'll die" (quoted in Puig de la Bellacasa 2010, 161).

We can only properly make sense of how an intimately shared world gives rise to an overlapping series of flexible limits and possibilities by expanding the territory of the commons, by placing the activity of lobster fishing within the wider social and ecological context that operates in a place like Castletownbere. By focusing on specific resources and the people who exploit those resources, we can get lost in the same abstract, bioeconomic framing that tends to dominate approaches to resource management. In doing this we ignore the *circulation* of the commons, the continuous ways in which a diversity of social and material resources are mobilized through commoning. From this perspective, we begin to see a world that is not oriented around the production and management of scarcity, but a world that relies on and reproduces diversity as a necessary condition for the commons. Understanding the commons as a flow of diverse resources and capacities between humans and nonhumans challenges the idea that simply *limiting* access to a resource is the best way of protecting it. It also pushes us into the middle of the drama, the "theater within which the life of the community is enacted and made evident," where the messy give-and-take of commoning unfolds in response to practical material questions (Hyde 2010, 31).

Circulation of the Commons

When I first arrived in Castletownbere on the Beara Peninsula, I barely knew anyone there or anything about it. The place existed for me only on maps and application forms for funding. During my time there, I grew attached to it in ways I could not have imagined beforehand. There was nothing spectacular about this process, nothing mysterious. These attachments developed out of a growing familiarity with the particular place I lived—the road I drove or walked down nearly every day, the pub I drank in, the view that I could see out of my window. Most important of all for thickening this relationship to place were my relationships with other people there and, more specifically, the things we did together. My neighbors, for example, lived just up the road and had lived in the area all their lives. They told me stories about different features of the landscape and the people who lived in the nearby houses. They also occasionally asked for my help moving cattle from one field to another or chopping and collecting wood. A few months after I moved in, we decided to grow vegetables in one part of their field. My neighbor Sam had told me about how his family used to grow vegetables when he was small. Sam and I dug and planted the garden together. Over the subsequent spring and summer, we used to meet in the garden; it was something we shared and did together. Other encounters and experiences emerged from this arrangement. I wanted to get some chickens, so he helped me build a chicken run with his tools and materials as well as branches we collected from the forest.

One day Clare, Sam's wife, met a local fisherman and neighbor in the shop. After she told him about me, he said he would invite me to go fishing. A few days later, I dropped around to his house, which was just down the road. We spent the evening chatting, and at the end he said he would give me a ring when a place opened up on a boat. A week or so later he called me.

I had been in Castletownbere for six months and had not managed to get out on a trawler. According to a retired fisherman who now worked for BIM, no fisherman would trust me as they would think I worked for the government and was trying to catch them out or that my research would be used against them. When I tried to go through the "official" channels and get a "research" position on a boat, I was told there were already many "observers" who wanted to go out for scientific research and I would not get a place. I was also told I would need to do a three-day sea survival course, a statutory requirement, and procure a lifejacket and steel-cap boots. Johnny, the fisherman I met through my neighbors, lent me a lifejacket and oilskins before I went out on the boat. I was also not going as an observer but to work.

This is an incidental moment but it illustrates how things like getting out on a trawler happened through everyday, uneventful exchanges with people, exchanges that were not just social but also material. It was only when I began living in the place, relating to the people who lived there other than as "research subjects," that possibilities presented themselves and unexpected encounters emerged. If I had not been living in Castletownbere, if I had not been engaging with my neighbors, walking the roads, and digging the garden, then it would have been very hard for me to meet the fishermen that I did. In a town like Castletownbere, and on the peninsula more generally, the circulation of information and resources through these informal circuits and social networks was vital to the ongoing existence of the place. It was only because of these everyday encounters and forms of exchange that things were able to happen. The material and immaterial mesh together as social relationships forged through encounter and activity enable the circulation of resources and information, which in turn constitute relationships. There is a long history of this kind of social economy in a place like Castletownbere because it is hard to survive without the help of other people and without access to a myriad of social and material resources.

Most fishermen in Ireland are not paid a wage but instead depend on income from what they catch. In the inshore sector, most of the boats are owned and skippered by the same person. This means they are chronically insecure and precarious because they do not make any income unless they can catch *and* sell their fish, a situation that pincers them between the vagaries of the sea and the market. But it also means that these individuals directly control their own means of production and labor. They are not directed or managed by a boss or the clock. It means they can, and must, rely on other forms of economic activity outside the wage and commodity-market. It means they must be inventive and attentive to the diversity of use-values and nonmonetary exchange that exists around them. While crewing on a trawler mostly required repetitive, physically demanding labor, inshore fishing on your own boat demands a whole repertoire of skills, including boat repair, crew management, identification of the best fishing spots, strategies out at sea, and so on. These different skills mean not only that a fisherman is more self-reliant but also that the work is more interesting and rewarding. While the unpredictability of fishing can be, and often is, a frustration to fishermen whose livelihoods depend on the size and quality of their catch, there is also an explicit awareness of how this unpredictability "makes things interesting."

Joe makes most of his income from lobster because they are the most valuable fish, but he also targets crab with the same pots as well as flatfish and mackerel with his nets. He values the flexibility this brings. It allows him to shift between different fishing grounds and species depending on changes in the weather or movements of the fish. For example, lobster pots tend to be dropped closer to shore. This means that if the weather is bad it is not always possible to shoot for lobster or haul pots that have already been shot. If Joe knows that the weather will turn bad, then he is better off shooting his pots further from the shore to target crabs. While having a boat fitted out with nets makes him less mobile in the water, potentially preventing him from accessing other fishing grounds, it opens up other possibilities in terms of what he is able to catch. Most importantly, the nets make him almost completely self-sufficient for bait. After fuel, bait is the most significant cost for crab and lobster fishermen. Nearly all the potting fishermen I met had devised ways of accessing free or cheap bait. This involved keeping the dogfish and conger eel that appeared in the pots, using by-catch from nets, or else coming to some arrangement with other fishermen they knew who went trawling. Tom, for example, got undersized mackerel off his son's boat and kept it salted in an abandoned hold by the fish factory over the year. He was thus able to make a living from lobster fishing not simply by extracting and selling lobsters but also through social and material resources he was able to mobilize on land and sea that are not reducible to particular biophysical stocks or "things." They are embedded within, and inseparable from, the social relations and networks that maintain them.

Sociologist Maria Mies has written about the importance of this localized circulation of production and consumption as an alternative way of understanding waste. Although capitalist modes of production tend to separate the point of production from the point of consumption, subsistence and commons-based economies incorporate the two, transforming "waste" into something productive and reproductive. This is because the value and wealth of the commons are not reducible to exchange-value and commodities. As Mies concludes, "Production processes will be oriented towards the satisfaction of needs of concrete local or regional communities and not towards the artificially created demand of an anonymous world market. In such an economy the concept of waste, for example, does not really exist" (2001, 1011). The redefinition and recirculation of "waste" as surplus is what ensures the productivity of the commons, generating a wealth of use-values rather than a limited number of commodities with market value.

While Mies focuses on the ways waste recycles within the commons as an immediate use-value, it is also the case that resources and labor can be invested in the commons that do not have either an immediate use-value or exchange-value. This opens up a temporal aspect to the (re)production of the commons that illustrates the entanglement of material practices and the subjectivities and social relations they require and constitute.

At the beginning of my research, I worked harvesting mussels on a farm for a few weeks. The mussel farmer, Frank, had been a fisherman but had shifted occupation because he did not see much of a future in it. At the end of each day, we would leave the harvested mussels at a small pier for collection by the wholesaler, who was also his neighbor. One day Frank put the harvested mussels on his trailer and dropped them at the bigger pier so they would be easier for the wholesaler to pick up. When I asked him why he did it he said, "You just never know, I might need his help some time if I break down or whatever." Frank also always left a bag of mussels hanging from the pier where he landed his boat. They were mussels he collected from the deck of his boat that by law are supposed to be discarded at sea. He left them there for neighbors and friends and told me if I ever wanted mussels I could come by and take them.

Frank did not calculate these gifts of time, labor, and resources in simple utilitarian terms, nor were they acts of altruism. They were the actions of someone who knew intimately and immediately that they were part of a wider collective on which he relied. Things can always *become* useful or valuable by being invested somewhere else. You never know when conditions will change and something may be needed or brought into play. The refrain "you never know" corresponds once again with a tangible awareness of how unpredictable things are and how the continuing support of others allows you to deal with that. When Tom's tangle nets had been out for five days because of the bad weather, he called me and his friend to come out and help him because he could not manage them himself in the rough weather. In December, with the prawn fishing starting, he called us both up again to help him take down the pots from a pier where he had kept them over the summer. Such work cannot be done alone, but it is also difficult to hire someone for the day, especially when it might be at short notice to coincide with the tides or good weather. Help does not come easily, nor is it a result of fixed actions or a form of membership. Participation in this wider collective is an ongoing, everyday process that almost recedes out of sight when it is reproduced through such minor acts of gift-giving.

In response to chronic uncertainty and the limited capacities of any one individual to deal with this uncertainty, the commons represents a collective reserve of social and material investments. The many gifts exchanged on a daily basis in a place like Castletownbere represent a social surplus, a promise for the future, which is not limited to a calculable transaction between two individuals. The work of the writer Georges Bataille and his concept of the "general economy" is instructive in thinking about this surplus and the need to conceptualize the commons outside the terms of liberal economics (Bataille 1991). Unlike liberal economists analyzing economy from the perspective of production and the management of scarcity, Bataille approached the economy from the perspective of consumption of wealth and the management of surplus. His economic anthropology of different cultures and civilizations examines how surplus wealth is socialized through forms of ritualistic destruction, lavish consumption, accumulation, and war. He did not, however, examine subsistence economies where the expression "I store my meat in the belly of my brother" carries such significance. This expression conveys the simple idea that by gifting surplus resources or time to a neighbor, one is effectively investing in the future. In this sense, where the care-practices involved in commoning are about recognizing and negotiating material limits that inhere in any context, the practices involved in sharing and circulating surplus relate to the exchange of gifts and the negotiation of nonmonetary wealth.

Describing similar practices of reciprocity, mutualism, and gift-giving in the making and re-making of community, theorist Massimo De Angelis writes,

> What in the Andes is called ajnji is *a form of reciprocal labor that is the weaving of the social fabric of a community* (or Ayullo in the Andes) through circuits of reciprocity, and it is based on principles of often implicit and not announced or bargained equality matching between individuals or community: today you do this for me and tomorrow

> I'll do this for you: a kind of circular "gift economy" as discussed by Mauss. (De Angelis 2007, 176; emphasis added)

De Angelis argues that such sharing is often characterized by conviviality. This conviviality confers on the labor something more than just a way of saving on costs. Nearly every time I finished working on a boat, I was encouraged to take as much fish as I could carry. These qualities of conviviality and generosity, however, are matched by resentment toward those who do not reciprocate and participate in these forms of sociality. I learned that fishermen once used to give away fish to people who would come down to the quay. They stopped when someone began asking all the boats for some fish and then selling it. The image of harmonious gift-giving and reciprocity must also be contrasted with conflict and disagreement that emerge when informal, socially accepted and understood boundaries are crossed.

The Tragedy of the Commons

Although better regulation of the inshore fisheries was something Joe wanted, he was clear about where the real problem lay: the price of fish. It was not viable anymore. It did not matter how healthy or how well regulated the stock was—a fisherman could not survive off his income. At the time, the price of lobster was 9.50 euro per kilogram instead of 13 euro per kg the year before, and brown crab was 1.05 euro per kg rather than 3 or 4 euro per kg. Joe said he was competing against cheaper imports of worse quality fish as well as retailers and processors who did not pass on any extra money. Because of these market pressures, fishermen were often being forced to overfish.

The first day I went potting with Joe, the processing factory had told him they were not accepting the bodies of crabs because they could not sell them. The processing factory had been in decline, and Joe was worried he would have nowhere to sell his catch—or that he would have to sell it to the processing company in Castletownbere, which would involve landing the catch farther down the coast. He had no choice but to rip the claws off all the crabs caught—maybe two to three hundred. It is illegal to rip both claws off the crab because without them they have no way of hunting, essentially ensuring they die. Joe was also aware of how painful it must be for the animal. Joe knew all this but had no choice. Not only did he run the risk of a fine and a criminal record if he was caught, but he also thought it was a "terrible waste of good meat." In addition to dealing with the "exogenous violence of the market," Joe was also competing with lobster boats that were organized around different priorities (Caffentzis 2004).

At the time of my research, the local processing factory owned five boats in the Castletownbere fleet. Although these boats did not look particularly different from other inshore boats in terms of size, they are specifically designed as potting boats and are thus more commercially oriented: they did not have any excess gear, such as heavy nets, so they were lighter and faster in the water, and they had larger engines so they could move faster between fishing grounds. These boats were thus able to cover greater distances and shoot more pots than a smaller boat could: 800 pots compared to the average 500.

Brian, the skipper of one of the factory-owned boats, told me that if the factory had its way he would be out every day checking the pots and shooting them again—something that would be highly destructive for the crab and lobster stocks. This was not possible, however, because of the cost of petrol. He was already spending 150 euros a day compared to the 30 euros spent by fishermen like Joe. Because Brian did not have nets, he was also unable to catch his own bait, something that the company took advantage of: they sold bait to Brian at a cost of 60 or 70 euros per day. The boat itself was divided in seven shares (rather than four), with three of those going to the owners of the boat. This meant that for the same amount of work—a day's lobster fishing—Brian made half as much money as Joe. It was no surprise that Brian was doing up his own boat on his days off in the hope that he would fish for himself in the future.

By fitting out and equipping their boats specifically for crab and lobster potting, the factory-owned boats were operating according to the logic of the market. At the same time, factory ownership of these boats was turning the fishermen who operated them into wage earners rather than independent operators with control over the productive process and a stake in the resource they depended on. While Brian was not paid a wage exactly, his activity was valued in terms of how many crabs he caught, not how he caught them. The flip side of this is that it becomes harder for independent fishermen like Tom or Joe to operate when their livelihood is being undermined by more "efficient" operators whose value practices are determined by the commodity market. The situated relationships, practices, and values associated with the commons must always be understood in the context of the "exogenous violence of the market" and the liberal responses that attempt to resolve such contradictions through new forms of regulation and exclusion (Caffentzis 2004).

These contradictions are tangible for most fishermen, particularly in the inshore fisheries where many of the fishermen I met are caught between the pressures of the market and environmental regulations that limit possibilities for commoning. For example, the ban on discards across the fisheries has meant that it is technically illegal for Joe to collect the assorted flatfish in his nets and use them as bait for his pots. The same restrictions apply to the dogfish and conger eel that appear in the crab and lobster pots, or the mussels that are left on the deck of the boat after harvesting. Clearly, the discarding of vast amounts of fish is something that must be stopped, but the ban on discarding *per se* ignores how by-catch can operate according to a different social, ecological, and economic logic—as part of the circulation of surplus within the commons, rather than the production of waste.[23]

My argument here about the inshore fisheries has been made many times in different contexts by political ecologists since the mid-1980s (Blaikie 1985; Peet and Watts 2004). Put simply, the degradation of natural resources and environments are not simply the fault of "irresponsible" resource users but need to be understood within the broader, historical context of capitalist modes of production and biopolitical "improvement." This contextualization does not just provide more critical narratives for understanding environmental degradation, but also identifies different ways of producing and organizing natural resources. In this way, political ecologists have also provided important counternarratives by acknowledging the historical and continuing persistence of different relations to nonhuman nature, such as those I have described in this chapter (Schneider and McMichael 2010). The difficulty with this work is that these situated ecologies of value and knowledge are often hard to see and describe.

Referring to the "silence of the commons," philosopher Ivan Illich writes: "the law establishing the commons was unwritten, not only because people did not care to write it down, but because what it protected was a reality much too complex to fit into paragraphs" (Illich 1983, 6). These lines should not be read as a romanticization of a premodern oral culture but rather as an observation about the practical, messy, and negotiated nature of the commons that persists today. Illich's emphasis on improvisation and pragmatism is echoed in Linebaugh's suggestion that "commoners think first not of title deeds, but of human deeds: How will this land be tilled? Does it require manuring? What grows there? They begin to explore" (Linebaugh 2008, 45). The activity of commoning emerges from the ground, in the midst of the ongoing, everyday interactions and exchanges that take place between, and constitute, those who take part. Significantly, the commoning described here is not the outcome of any self-conscious, collective resistance to the market or the state. Historic and contemporary forms of commoning on the western seaboard of Ireland have mostly arisen out of pragmatic, material responses to particular needs and problems. These problems are clearly exacerbated by the unpredictable marine environment and the historic lack of financial and technical resources, both of which lead to relations of mutualism and reciprocity that have enabled social life to go on. "The greater the need to improvise," anthropologist David Graeber writes, "the more democratic the cooperation tends to become" (2011, 96). As a consequence, however, the commons always appears to recede into the background, to be recalled only after it has disappeared. This is the real meaning of the tragedy of the commons, one that needs to be challenged through the rediscovery and novel articulation of those subjectivities, forms of knowledge, and value that are not just about "local" communities but about different ways of organizing the material dimensions of life. These collective and shared ways of relating to one another are not just spontaneous but require ways of knowing and doing that in themselves constitute the subjectivities and natures on which the commons depends.

However, in thinking about the "invisibility" of the commons and the need to recall and reclaim them, it is also important not to romanticize or idealize them, or to ignore how collective forms of organization and production are being coopted within new liberal regulatory frameworks and new modes of capitalist production (McCarthy 2005). Writing about his own work mapping the commoning practices of fishermen in New England, for example, Kevin St. Martin asks, "What economic potentials do we forfeit by seeing the fisheries commons of New England as always a location where community and commons-based economies are retreating and capitalism is advancing?" (2005, 67). This emphasis on "economic potentials" represents an affirmation and valorization of commons-based economies that must be tempered by a critical appreciation of how these capacities for self-management have become the target of state policies and programs keen to shift responsibility down to communities (De Angelis 2013). St. Martin draws on the influential work of the collaborators J. K. Gibson-Graham (Julie Graham and Katherine Gibson), two geographers whose conceptualization of a postcapitalist politics attempts to weaken the hold of capitalism on the imagination and open the potential for different future trajectories. While this is an important and necessary task, such an approach can also tend to set up too fine a distinction between capitalist and noncapitalist economies and subjectivities. As feminist and Marxist scholars have argued for some time, the sphere of reproduction (and commoning) has historically been the invisible but *necessary* condition for capitalist wage labor and accumulation (Federici 2004; Moore 2014a, 2014b). Today, this relationship is changing again as the sphere of reproduction and the commons has become the target of new forms of commodification and control. Efforts to render commoning and the commons more "visible" should always acknowledge the creativity of capitalist enclosure and liberal governmentality.

6

Conclusion

Neoliberalism and the Commons

Why Malthus Still Matters

In his recent intellectual biography of Thomas Malthus (2014), Jonathan Mayhew argues that his 1798 *Essay on the Principle of Population* was as much a response to the radical political treatises and upheavals taking place across Western Europe in the 1790s as it was a reaction to the problem of overpopulation. Malthus was opposed to what he perceived as the naive and dangerous optimism of revolutionary Romanticism and Enlightenment progress. In Malthus's view, these cultural, political, and scientific movements shared a critique of existing institutions and a utopian belief in human perfectibility and social progress.

Malthus had a particular opponent in William Godwin, the husband of Mary Wollstonecraft and a political philosopher who wrote *An Enquiry Concerning Political Justice* in 1793, when the French Revolution was in full swing. Godwin took up themes that were common to Rousseau and Thomas Paine. He argued that there was no predetermined human nature and no propensity to evil, in which case the causes of pain and suffering must be traced to social conditions, the type of government, and their corrupting influence. The answer was to destroy these malign institutions so as to liberate humankind, replacing the artificial production of scarcity with natural abundance. In the preface to the first edition of the *Essay on Population*, Malthus writes, "Let us imagine for a moment Mr. Godwin's beautiful system of equality realised in its utmost purity." Under such conditions, the population would rise, with the consequence that "the corn is plucked before it is ripe or secreted in unfair proportions; and the whole black train of vices that belong to falsehood are immediately generated" (Malthus 2013).

Mayhew draws a picture of a man who was inspired to write his 1798 essay because of the growing support for such "dangerous" ideas, ideas that supposedly had no basis in the social and material realities of the time. Although writers such as Godwin professed a reverence for reason, they lived, Malthus wrote, in a world of delusion, a "fantasy of reason." Nature was not a Garden of Eden that simply had to be liberated from corrupt institutions, nor was it fully amenable to human control and desire. Nature was first and foremost a complex interplay of environmental and socioeconomic forces that unfolded outside any idealistic aspiration of man. Questions concerning the optimal organization of society had to be asked in respect to the specific conditions that existed. Malthus concluded that misery and scarcity should not be understood in moral terms, as good or bad, but rather as *real* phenomena that were amenable to analysis and management; any realistic progress in human society had to be worked out with regard to "nature as it really was."

Clearly sympathetic to the "down-to-earth" realism of his subject, Mayhew writes, "Malthus was the untimely prophet of how population, economy, and the environment interact; from each of these directions, our present global predicaments seem to demand that we attend to his insights and to the responses they have garnered as we plot our path in troubled times" (2014, 4). Although Malthus has been identified as the proponent of savage social policies, his arguments, Mayhew argues, were based on a close analysis of the available empirical and statistical data that existed.[1] Through a close reading of Malthus and his work, Mayhew attempts to rescue him from a superficial analysis that reduces him to a crude defender of class interests or advocate of repressive population control. He concludes his book: "Malthus should be read rather than merely caricatured, and he should be read not though a 'right-wrong' binary of timeless truth but for his temper of mind. *In other words, where Malthus has normally been read in scriptural terms as right or wrong, he is better read as an 'organon,' a tool by which to reason*" (Mayhew 2014, 235; emphasis added).

It is not necessary to agree with Mayhew's conclusions to recognize the validity of this statement. Understanding Malthus as an "organon" of thought can help us decipher what is at stake in neoliberal environmentalism today: a "temper of mind" that absents questions of right or wrong in favor of an assumed "down-to-earth" economic analysis. As theorist Thomas Lemke writes regarding this form of reasoning, "The coordinates of governmental action are no longer legitimacy or illegitimacy but success or failure" (Lemke 2011b, 46). As I have shown throughout this book, the different forms of governmentality that have emerged in response to the problem of overfishing all revolve around a common desire to *better account* for the underlying causes of the problem: how can we intervene in a way that does not deny realities on the ground but instead harnesses and mobilizes them to preserve marine resources and allow economic growth?[2] This formulation of the problem succeeds in turning overfishing into a technical problem that effectively naturalizes the particular capitalist relations of production that have caused widespread degradation of marine resources. It also elides the ongoing existence and potential for alternative ways of relating to, using, and valuing marine resources. At the same time, this pragmatic, seemingly neutral approach to the problem of overfishing has dramatic and unprecedented effects on the fisheries and, most directly, the fishermen whose livelihoods depend on access to those fisheries. In the first three chapters, I sought to trace out the contours of these transformations and the different modes of liberal governance they correspond to.

In the first chapter, I identified how the crisis of overfishing was being framed in familiar liberal or Malthusian terms as a problem of scarcity. The recent Common Fisheries Policy addresses the need to preserve biological fish stocks at sustainable levels by bringing fishing effort into line with their reproductive cycle. This approach is crystallized in the goal of Maximum Sustainable Yield. While it provides a shared, bioeconomic goal for those involved in the fisheries, the pursuit of MSY has provoked critical reflections on the nature of fisheries management and the limitations of an approach that fails to acknowledge and account for the unpredictable behavior of fish and the economic self-interest of fishermen. Rather than denying the bioeconomic "nature" of the fisheries, these elements become the target of new analysis and intervention, with the validity of these interventions being assessed in terms of how *effective* they are in bringing fishing effort into line with fish stocks. The assumed character of the market in this case is very much in keeping with the liberal belief in a self-regulating sphere of exchange operating beyond the state, an assumption that gave rise to laissez-faire policy-making. Instead of preventing fishermen from acting in their own self-interest, the liberalization of the fisheries involves harnessing this self-interest by effectively rewarding more efficient use of resources through the market. Liberal reforms have ultimately resulted in the ban on discards and the introduction of individual transferable quotas, understood as the only way of returning to a "virtuous cycle"—the "natural" balance of demand and supply.

Where the "natural" limits of fish stocks and the economic behavior of fishermen provided the rationale for the ban on discards and the liberalization of the quota system, it is the promise of the "green" economy that provides the basis for the specifically *neoliberal* governmentality examined in chapter 2. Although the basic issue being addressed is still the environmental and economic impasse of industrial-scale fishing, the neoliberal response focuses not on biological or economic limits but rather on the construction of new opportunities through measurements of environmental performance. In this case the "free" market is actually part of the problem because global competition has forced the Irish fishing industry to overexploit limited fish stocks, even while falling behind in the race to profitability. Significantly, however, the market is not jettisoned as a basis for good governance but is rather extended and transformed. Rather than simply allowing *the so-called free market to function*, the neoliberal approach to policy making actively cultivates new markets in order to generate new forms of environmental entrepreneurialism and competition. By combining environmental concerns with the new market for "sustainable" seafood, this mode of neoliberal governmentality also shifts the generation of value from the extraction of fish to the capacity to represent one's environmental performance in the marketplace. In effect, marine resources are no longer understood to be the main commodity. Rather it is the informational content that identifies and differentiates the fish product in global markets. The process through which this information is produced, however, is highly technical and uneven, benefiting those in the industry who are better able to play the game rather than reflecting or addressing the situated social and material needs of particular fisheries. Despite the rhetoric of greater transparency and openness, the introduction of auditing technologies like the Environmental Management System and associated accreditation processes reveal a more intimate role for state agencies and a highly bureaucratic and opaque governance network of unaccountable private intermediaries and nonstate actors.

Assertions that the "state is retreating" or that the "market is advancing" assume that the state and market are discrete, static entities rather than partners adapting their institutional forms and roles to specific contexts and problems. The benefit of this analytic perspective is illustrated in chapter 3, where support for community-based resource management is identified in scholarly and policy literature as an alternative to both privatization (market) and state-control (public). This form of governance is novel in so far as it harnesses the "community" as the instrument for achieving environmental and economic goals, a community that is constructed through the drawing of "soft" territorial boundaries and access rights. At the same time, this approach formulates the overexploitation of lobster stocks in classically liberal terms, accepting it as the outcome of underlying economic and biological tendencies that must be accommodated within adaptive institutional frameworks. This example, more than any other perhaps, reveals the commitment to "making policies work on the ground," seeking at all times to construct institutional parameters that are able to harness and coordinate the activities of fishermen and their interactions toward the narrowly defined goals of biological productivity and economic profitability. The degree of time and work committed to this task reveals both the reflexive and constructive character of liberal governmentality. The institutionalization of the commons can, in this case, be seen as a novel advance in the form of the liberal state: the "supplying" of "collectives" with the possibility of actively participating in the solution of specific matters and problems which had hitherto been the domain of specialised state agencies specifically empowered to undertake such tasks" (Lemke 2001, 202).

The move toward community-based resource management is framed within a discourse of empowerment, participation, and responsibility that threads throughout the first three chapters. Significantly, this "transfer" of freedom or agency to individual fishermen is not illusory even if it is considerably delimited within economic and technomanagerial spheres. Fishermen are no longer the passive targets of top-down regulations; the ban on discards, ITQs, eco-labels, and community-managed fisheries all offer opportunities to fishermen who are willing and, much more importantly, able to take them. The carving out of these opportunities represents a complex process of inclusion and exclusion that is not just imposed "from above" but worked out through the iterative crafting of normative parameters that both assume and construct new subjects and natures. The function of liberal forms of government is to generate these fields of productive possibility within which individual actions and trajectories are both constrained and enabled. This aspect of liberalism can be obscured when analyses fail to situate it historically and trace its capacity to respond to and absorb obstacles and critiques. As illustrated in the first three chapters, efforts to limit fishing effort or create new opportunities for fishermen were open-ended. The persistence and flexibility of these efforts owed not to a blind faith in the market but a far more pragmatic concern with changing the behavior of fishermen in order to conserve fish stocks and generate new sources of profit within an emerging market for "green" seafood. Along these lines, there are no contradictions between "ideal" forms of (neo)liberalism and their "actual" realization in particular spheres of society. It is precisely the messy, contested, and reflexive nature of (neo)liberalization that makes it such a dominant force crafting our ecological futures. The lack of engagement by fishermen with the Environmental Management System was not understood, for example, as a problem with the idea (incentivizing fishermen to adopt "green" practices in their operations), but a failure to make the incentives effective or real enough. This in turn provokes the need for further investments of time and effort until the initiative does work because fishermen will presumably have no choice but to participate.

Current efforts to fix the problems of "first modernity"—a mode of production and regulation that understood and valued nature as raw material for commodity production—are carving out *new* ways of understanding and valuing nature that in turn open up new opportunities for capitalist accumulation (Bakker 2010). Capitalism alone does not overcome these ecological limits by simply enclosing and commodifying new areas of life. It is facilitated in this process through biopolitical transformations that take as their point of departure the need to preserve and conserve "nature" by absorbing it into the bioeconomic calculations of government. We cannot fully understand the ongoing response to ecological crisis if we see it only as a crisis of capitalism and not as a crisis of government, a new series of questions that challenge policy makers, scientists, and resource managers to transform how we understand and manage our relations to nature. As is evident with the management of the fisheries, these new modes of government cultivate, and render invisible, certain ways for fishermen to relate to one another and their marine environment. This does not mean we have to accept these economic rationalities as all-dominant, but it does mean accepting them as something more than an elite, capitalist project that can be "revealed" through the correct analysis.

This perspective brings us back to Malthus's critique of Godwin: a world of abundance is not just waiting for us if we can only "pull the scales from our eyes" and tear down the corrupt institutions that stand in our way. Godwin's idealism, while attractive, loses sight of the reality of our current socio-natural order.[3] As Dardot and Laval write, "To go on believing that neo-liberalism can be reduced to a mere 'ideology,' a 'belief,' a 'mind set' which the objective facts, duly registered, would be sufficient to dissolve, just as the sun dispels morning clouds, is in fact to mistake the enemy and condemn oneself to impotence" (Dardot and Laval 2013, 19). This is what Foucault and Marx help us to understand. The reality we inhabit is not an "illusion" or façade; it consists of real social, economic, and biophysical forces that shape our lives, now more than ever. We are all entangled within institutional and economic practices that shape not only the way we think but also the way we act, the way we work on ourselves and the people, things, and places around us. "Neoliberalism," writes literary critic Steven Shaviro, "has entered into all our preconscious assumptions; it permeates our habits of thought and speech. Even when we seek to oppose the most outrageous depredations of human livelihoods and of the physical environment, we find ourselves using the language and the presuppositions of cost-benefit analysis, optimization, and so on" (2011, 79). In this context it is difficult to imagine, let alone enact, a different world, one that is not enmeshed within the rationalities and practices of (neo)liberal capitalism and the scarcity it (re)produces for the majority. There is not going to be a moment of realization, no "storming of the Bastille," as Ulrich Beck calls it, that will resolve the ecological catastrophe of the present (Beck 2010). Any hope of alternative ecological futures will emerge not just from ideological critiques of neoliberal natures but from the material remaking of new collective subjectivities, which brings us to the more-than-human commons and commoning.

Commoning Toward an Alter-Biopolitics

"Nature," "environment," and "community" were not terms that many people I met in Castletownbere used, yet their everyday lives were grounded in and distributed across this place and those they shared it with. Castletownbere was neither a pristine area of scenic beauty nor "just a place to work," somewhere taken for granted by those who were "too busy" to look up from their tractor wheels or fishing nets. It was a place that was conjured up through banal, everyday encounters and exchanges that allowed people to go on making a living through recourse to vital social and material resources. This emphasis on the "invisible" but palpable, the unrepresentable but tangible, not only resonated with my research experience but also with a familiar insight often made about the commons: that the wealth of social relations and practices that are concerned above all with the immediate, taken-for-granted matter of sustaining life are often silent and ignored. In other words, the difficulty of identifying and describing the activity of commoning was not just a methodological problem but something that gets to the heart of the real tragedy of the commons.

In chapter 4, I shifted attention to the all-too-often-invisible world of commoning and sought to differentiate it from the liberal or institutional commons described in chapter 3. A key argument here was that the institutional commons fails to escape the liberal underpinnings of the "tragedy of the commons" narrative: the need to remedy the "distortions" of unregulated access to a resource, to transform it into a "common property regime," is predicated on the bifurcated world of the liberal economic subject and the limited stock of biophysical resources. This division leads to the implanting of territorial boundaries, access rights, and new relations of governance that operate on the basis of individualized, economic subjectivity and nature as resource stock (Goldman 2004). In contrast, the more-than-human commons can only be seen or made sense of if we assume that *the world is already shared.* While fishermen, for example, are represented as self-interested economic actors, this is only part of the story, the part that is targeted through forms of liberal governance. The other part of the story begins with a subject who is situated between, and reliant on, other humans and nonhumans. This social and material interdependence gives rise to practices of care and reciprocity that do not just enable but require that resources and capacities be shared. Commoning thus operates within limits—not absolute limits but the social and material limits that inhere in any context where things are shared. It is the ongoing negotiation of these limits that generates collective capacities and constitutes the commons. I want to finish by making two general points about the concept of the more-than-human commons and how it can help us to think beyond the liberal, humanist assumptions that underlie Malthusian and post-Malthusian accounts of ecological crises and social change.

I have already shown the limitations of the liberal or institutional perspective on commons. There have also, however, been attempts to identify the emergence of new "social" or "immaterial" commons, such as the knowledge and cultural commons , the digital commons and peer-to-peer production, and the biopolitical commons (Hyde 2010; Bauwens 2005; Hardt and Negri 2009). Although the political perspectives that inform these analyses differ, they all assume an analytic distinction between the "social" commons and the "natural" commons that map on to the historical distinction between liberal and neoliberal political economy and their corresponding modes of industrial and postindustrial capitalist production.

In his article "Two Faces of the Apocalypse," for example, theorist Michael Hardt describes the difference between anticapitalist activists and climate change activists at the United Nations Climate Change Conference (COP 15) in Copenhagen (Hardt 2010). While the first group insists "another world is possible," the second adopts the slogan, "There is no Planet B." Hardt traces these different political positions back to their contrasting notions of the commons. On the one hand, anticapitalists consider the commons a social product, representing the wealth of human labor and creativity. On the other, environmental activists identify the commons with natural resources, including the atmosphere, rivers, forests, and biodiversity. Hardt argues that the "natural" commons is subject to the logic of scarcity, bringing us into the domain of liberal economics and the institutions of formal and informal property rights, as illustrated by the work of Elinor Ostrom and other economists discussed in chapter 3. In contrast, the "social" commons is not subject to the logic of scarcity; it is malleable and infinitely reproducible. The problem with this distinction is that we end up with one form of the commons that appears to be *asocial* (excluding the "thick" social relations and forms of knowledge and labor involved in *caring* for the "natural" resources that communities rely on), and another that appears to be *anatural* (excluding the material limits and properties of more-than-human bodies involved in the [re]production of complex life-processes).

Although the distinction between the material/natural commons and the immaterial/social commons can be helpful analytically, it tends to be overstated, obscuring the continuity and inseparability of the material and the immaterial, the natural and the social, when we consider the spheres of production and reproduction *together*. And while there are clearly profound new technical possibilities for the making and remaking of socio-natures—ones that take us beyond the "natural" limits of Malthusian prognoses—these can too easily be interpreted as tools for escaping material limits and the urgent need to radically transform our relations with nonhuman others. As theorist Luigi Pellizzoni writes, "The neoliberal entrepreneurial agent looks similar to a god, since the full pliancy of materiality to human designs leads to depicting agency in terms of an ultimately unconstrained will" (2011, 800).

Melinda Cooper captures the tension between the "naturalized" limits of the liberal commons and the impossible promise of the neoliberal common when she writes, "It therefore becomes urgent to formulate a politics of ecological contestation *that is neither survivalist nor techno-utopian in its solutions*" (Cooper 2008, 50; emphasis added). We need to find different ways of articulating the relationship between limits and possibility, relationality and agency, human and non-human, that do not collapse back into humanist ways of thinking and doing politics. The concept of the more-than-human commons is about pushing our frameworks beyond the limited, dualistic understanding of commons as a resource and social change as an exclusively human affair. In this reading, the more-than-human commons provides a counterpoint not only to what anthropologist Arturo Escobar calls the "analytic of finitude," a "cultural order in which we are forever condemned to labor under the iron law of scarcity," but also to neoliberal fantasies of infinite growth that tend to ignore material questions of reproduction (1999, 6).

Reflecting on the surge of indigenous politics in Latin America over recent decades, anthropologist Marisol De la Cadeña describes how these new movements belie the familiar terms of Western political philosophy and political economy. Although they have arisen in opposition to capitalist accumulation processes, the indigenous response has not tended to fit within Western ontologies that identify humans as political subjects and nonhumans as apolitical (Latour 1993). "Digging a mountain to open a mine, drilling into the subsoil to find oil, and razing trees for timber may produce more than sheer environmental damage or economic growth," she writes. "These activities may translate into the violation of networks of emplacement that make life locally possible—and even into the destruction of place" (De la Cadeña 2010, 357). While De la Cadeña identifies the importance of place in these struggles, her account should not be read simply as a defense of the local community or resources against the incursions of global capital. The concept and practice of place within Andean cosmopolitics signify a fundamentally different relationship between humans and nonhumans that amounts to an alternative *form of life*. These movements have sought to bring "earth-beings" and "earth-practices" into the political sphere and insisted that relational ontologies be recognized rather than separated and rendered manageable within existing discursive frameworks and political practice. What is at stake in these new indigenous movements is not just access to resources or the value generated through their extraction but a radical challenge to the epistemological dichotomy of Nature and Humanity and thus a disagreement over the kinds of relations and practices that exist and "count" between collectives of humans and nonhumans (Ranciere 2008).

The sense in which Andean cosmopolitics represents not just a different claim to resources but a different world of relations between humans and nonhumans, a different organization of life processes, resonates with a long history of struggles against what De la Cadeña calls the "singular biopolitics of improvement" (De la Cadeña 2010, 346). Understood in this way, the more-than-human commons challenges not only capitalist commodification but also the liberal, dualistic epistemologies and knowledge practices that shape hegemonic forms of environmental regulation and control. At the same time, this conceptualization of the more-than-human commons does not mean idealizing subaltern or indigenous environmental practices and knowledge.[4] While the commons has a history and a historical continuity, it is not a fading premodern form of life or cosmology. The existence and appearance of forms of (re)production grounded in an immediate and intimate sense that the world is *shared* runs alongside and underneath dominant narratives of modernity and progress; it provides us with a countermodern *reference*, a "different praxis and management of the material dimension of life—in other words, an entirely different 'economics,' or way of ensuring the satisfactions of our material wants and needs," as philosopher Freya Mathews puts it (Mathews 1999, 130). It can always "arise unexpectedly in relationships among peoples and between people and place" (Rose 2004, 6). It should be possible to learn from the historical praxis of the commons without resorting to nostalgia.

This relates to the second point about the more-than-human commons: the commons is not a "thing" any more than the state, the market, or capital are "things." The more-than-human commons refers to the *relations* and *practices* that enable resources and capacities to be circulated and shared rather than accumulated, owned, or controlled. Following Linebaugh, it therefore makes more sense to talk about *commoning*, the ongoing social and productive activity that constitutes the commons. Understanding commoning as an activity that must always be demonstrated and practiced can help us avoid the tendency to essentialize the commons as something that already exists, or has existed, as a particular set of institutions or norms governing a community. The commons should not be relegated to some vestigial culture or local subsistence economy that has lingered on in geographically remote or "underdeveloped" parts of the world (St. Martin 2007). It is this kind of shrinking of possibilities that now feeds directly into new liberal modes of governance that responsibilize communities within an increasingly unequal and crisis-ridden global context. From this perspective, the conflation of the commons with local self-governance is more likely to reflect the state's response to mounting social, financial, and ecological problems. The uncritical tendency to localize the commons within particular communities and institutions can also fail to acknowledge or address the growing, *global* power and threat of corporate and financial actors and ecological phenomena such as climate change (Caffentzis 2010). Sociologist Michael Goldman summarizes this one-sided view in even starker terms: we are in danger of "casting a blind eye towards the destructive forces of capitalist expansion onto the commons and a broad smile that beams at the 'underskilled' local commoner who defies all odds by protecting the commons" (Goldman 1998: 21).

It is only through the creative elaboration and expansion of commoning that we can forge collective possibilities beyond circuits of capitalist enclosure and biopolitical "improvement" (Dardot and Laval 2013; Dyer-Witherford 2006).[5] This process necessarily takes place within the ecological, technological, and social conditions we find ourselves in. It is in this context (not an imagined past or future) that we need to *invent* new ways of commoning that enact ways of using and sharing resources and capacities that do not rely on forms of proprietorship.[6] As social theorist Massimo De Angelis writes,

> The process of social constitution of a reality beyond capitalism can only be the creation of, *the production of other dimensions of living*, of other modes of doing and relating, valuing and judging, and co-producing livelihoods. All the rest, regulations, reforms, "alternatives," the party, elections, social movements, "Europe" and even "revolution," are just words with no meaning if not taken back to the question of other dimensions of living. (De Angelis 2007, 1; emphasis added)

Sociologist Dimitris Papadopoulos uses the term "insurgent posthumanism" to describe a "move from enclosed and separated worlds governed by labour to the *making of ecological commons*" (2010a, 135). He recognizes a shift away from politics as primarily a matter of ideas and institutions, to politics as an embodied everyday practice: "a move from representation to embodiment"; and a move away from considering the human subject as the main actor of history-making. This conception of posthumanist politics offers a more immediate, material form of justice that is less concerned with seeking inclusion through mechanisms of representation than it is with constituting material justice in the here and now (Papadopoulos et al. 2008).[7] Bringing human and non-human together in this way orientates us away from struggles over future systems of equality and justice toward the ongoing, ever-present struggle for the conditions necessary for life to flourish; while there is no outside, there are ways of making and doing in the present where "we encounter a re-weaving of the social and the material *through the development of new practices, knowledges and technologies*" (Papadopoulos 2014, 77; emphasis added). In other words, the production of new ways of knowing and doing does not emerge from an *a priori* ideological position but unfolds as particular collectives of people, animals, plants, places, and technologies collectively struggle against everyday forms of injustice and exclusion. From this perspective, the construction of a politics of the common(s) is only going to emerge through a process of collective subjectivization, the making of new subjects and "values that are grounded in material practices for the reproduction of life and its needs" (De Angelis 2007, 32).[8] Echoing this expanded notion of collective politics, theorist Maria Puig de la Bellacasa writes that we need to find alternative ways of world-making that cultivate "power with" rather than "power over" (2010).

Thinking the commons as an experimental form of alter-biopolitics also opens up possibilities to read the past in terms of the present. It allows us to consider together the geographically, culturally, and historically distinct struggles over what ecologies count in the contested expansion of capitalist valorization and biopolitical control.[9] Foucault describes how the development of biopower in the eighteenth century precipitated new forms of opposition that demanded recognition and rights in the name of the body and of life. "Against this power," he writes, "the forces that resisted relied for support on the very thing it invested in, that is, on life and man as a living being.... *What was demanded and what served as an objective was life, understood as the basic needs, man's concrete essence, the realization of his potential, a plenitude of the possible*" (Foucault 1998, 144–45; emphasis added).[10] Something of this demand and objective is found in the "anti-enclosure" poetry of John Clare, a contemporary of Thomas Malthus and expressive voice of the more-than-human commons.

Writing at almost the same time as Malthus, the poet Clare experienced firsthand the dramatic changes brought on through the enclosure and "improvement" of England's rural landscape. Helpston, in Northamptonshire, was one of the last areas of England to experience systematic enclosure—at the turn of the nineteenth century. For Clare, who grew up roaming the unenclosed fields and commons that reflected the communal nature of social and economic life before enclosure, the fencing-in of the land struck him forcefully (Barrell 2010; Bate 2003). Clare contrasts the "immensity" of this "unenclosed" landscape with its shrinking through enclosure: "Fence now meets fence in owners' little bounds, / Of field and meadow, large as garden grounds, / In little parcels, little minds to please, / With men and flocks imprisoned, ill at ease" (Clare 2004, 168). Throughout his "anti-enclosure" poems, Clare reiterates this theme: enclosure shrinks the landscape and in the process strips the world of its multiple meanings, potential, and diversity: "The thorns are gone, the woodlark's song is hush, / Spring more resembles winter now than spring" (Clare 2003, 36). This literal and metaphoric "shrinking" of the world is central to Clare's resistance to enclosure and his defense of the manifold commons (Bresnihan 2013; Neeson 1996).

If we read John Clare's poems as the expression of something more profoundly particular than any place or time, we can reach a vision of the more-than-human commons that takes us away from the mutually reinforcing narratives of scarcity, enclosure, and capital. His poetry is not attuned to an ideal of "nature" or even just a particular geographic locality. It is more fully concerned with the immediate, everyday mattering of human and nonhuman life in all its messy, ecological complexity (Morton 2008b).[11] The commons, while always situated, can transcend the particular, reminding us that "commoning, the commons movement, is not an alternative economy but an alternative to the economy" (Esteva 2014: 149). Clare's poems sing across two centuries like an ongoing response to, and denial of, Malthus's metaphor of "nature's mighty feast" and the neoliberal fantasies of ecological modernization. If the ghost of Malthus haunts us two hundred years later, we should also make room for Clare in our imaginations of the future.

Notes

1. Introduction

1. At a public talk held in Dublin in October 2014, the co-chair of the IPCC Working Group III sought to explain to the audience what was at the root of the problem of climate change. Professor Ottmar Edenhofer argued that the atmosphere was a global commons that had been allowed to deteriorate because of unrestrained industrial activity. Assuming that the assembled audience would already be familiar with the concept, he referred explicitly to the "tragedy of the commons" without explaining it. According to Edenhofer, this tragedy would end only by introducing incentives to limit carbon emissions, and this was the rationale for supporting carbon quotas and carbon markets.

2. A year before the London exhibition, an Irish newspaper commented, "An oil painting of the Claddagh fishermen at home on their native strand may be a very pretty and picturesque sight, but we prefer to see the Claddagh and other fishermen manning large, well-built, and fully-equipped boats, and fleets constantly going out and returning with the rich harvests of the seam, which are inexhaustible" (*The Irish Builder,* January 1, 1883, quoted in De Courcy Ireland, *Ireland's Sea Fisheries,* 72).

3. For example, Ireland was slow to "follow" other European countries such as France, Spain, and Belgium in developing a deep-sea fishing industry. This was largely due to government policies that sought to protect the interests of inshore fishermen and fishing communities by restricting the development of trawling companies. In 1955 the minister for agriculture, James Dillon, said: "I will not let anyone start a big trawling company based in this country because I believe it would destroy the livelihood of the inshore fishermen" (Dillon, Dail Eireann debates, 11–12).

4. The Brussels-based coalition of environmental organizations called OCEAN2012, for example, receives funding from the Oak Foundation, the Tubney Charitable Trust, and the Pew Charitable Trust. OCEAN2012 has six full-time staff members operating in Brussels as well as money to pay for consultations and scientific research. They were established to lobby

the European Commission on issues concerning the reform of the Common Fisheries Policy in 2012. Commercial interests such as Unilever, which set up the international Marine Stewardship Council accreditation scheme for sustainable fish, also carry significant weight in fisheries management NGOs.

5. It is for this reason that the Marxist historical geographer Jason Moore has suggested that the era of the Anthropocene should more accurately be called the "Capitalocene."

6. This analysis resurfaces with the work of early political ecologists such as Michael Watts, Richard Peet, and Piers Blaikie. While their work provides a more rigorous understanding of the ecological effects of capitalist modes of production, they shared with Marxist political economists the need to provide a more historical and structural analysis of the various causes of resource degradation or pollution (Blaikie, *Political Economy of Soil Erosion*; Peet and Watts, *Liberation Ecologies*).

7. Two lengthy papers by Noel Castree provide a thorough overview and critical analysis of both the strengths and limitations of what he sees as a general trend within the field of geography: the analysis of neoliberalism as a differentiated practice that takes place within specific contexts, and a turn away from the underlying logic, as he calls it, that characterizes neoliberalism in general. See Castree, *Neoliberalising Nature: Processes, Effects, and Evaluations; and Castree, Neoliberalising Nature: The Logics of Deregulation and Reregulation*.

8. Governmentality studies has tended to focus on how biopolitical rationalities come to generate and modify subjectivity in the areas of health, biotechnology, sexuality, and work (see Burchell, *Liberal Government and Techniques of the Self*; Brockling et al., *Governmentality*; Rabinow and Rose, *Biopower Today*). Poststructural theory more generally has been used by scholars looking at environmental conflicts and critiquing the epistemological and discursive constructions of Nature inherent in Eurocentric projects of development (Escobar, *Encountering Development*, *Territories of Difference*; Agrawal, *Environmentality*; and Peet and Watts, *Liberation Ecologies*). Julie Guthman has sought to bring the concept/method of governmentality into her analysis of neoliberal environmentalism, but she does not situate it within Foucault's general framework of biopower and the historical variants of liberal/neoliberal political economy (Guthman, *Neoliberalism and the Making of Food Politics in California*). Luigi Pellizzoni identifies the distinction between liberal and neoliberal "natures" in his analysis of contemporary forms of environmental governance in a time of "disorder" (Pellizzoni, "Reflexive Modernisation and Beyond"). Recent work on neoliberal conservation and biodiversity offsetting has begun drawing together Marxist and Foucauldian analyses of neoliberalism, biopolitics, and capitalism in interesting ways (Büscher et al., *Towards a Synthesized Critique of Neoliberal Biodiversity Conservation*; Sullivan, *Banking Nature?*).

9. While Foucault relates this new vitalist power to the appearance of the human species, it is interesting to think about how nonhuman life and processes also became the object of new analysis and knowledge. This is made evident by the expansion of the new fields of botany, biology, geology, geography, and agronomy. More specifically, biopower opens up a new territory of intervention that is not so much spatial as it is ecological: the target of political economy was the relationship between vital processes in so far as these affected the overall productivity of the population.

10. The largest single "improvement" project in Britain at this time was the draining of the marshlands of East Anglia, Essex, and Kent, "work that transformed massive areas from miasmatic zones of shortened lives into the most productive parts of the nation, albeit with considerable social dislocation" (Mayhew, *Malthus*, 18).

11. The overlapping, expansive process of "improvement" and "enclosure" has been described by Nicholas Hildyard and others as a "compound process": "Because history's best-known examples of enclosure involved the fencing in of common pasture, enclosure is often reduced to a synonym for 'expropriation.' But enclosure involves more than land and fences, and implies more than simply privatization or takeover by the state. *It is a compound process which affects nature and culture, home and market, production and consumption, germination and harvest, birth, sickness and death.* It is a process to which no aspect of life or culture is immune" (Hildyard et al., *Reclaiming the Commons*; emphasis added). Understanding enclosure as an extensive "compound process" involving social and natural life—the spheres of production and reproduction—corresponds with the work of feminist and social historians who document the changing role, meaning and value of "nature" in the modern period (Merchant, *The Death of Nature*; Starhawk, *Appendix A*). Silvia Federici's book *Caliban and the Witch,* for example, sets out to reexamine the period of enclosure not simply as the historical moment of primitive accumulation through the seizure and fencing-in of land, but also the enclosing of women's bodies within the domestic sphere and the violent eradication of those knowledge practices performed by women relating to childbirth and healthcare. Federici identifies enclosure as the separation not only of people from the material resources they relied on but also of the spheres of production and reproduction, the latter being the sphere of women and the "natural," and the former being the sphere of men and capitalist valorization (wage-labor and commodity).

12. "The issue of government as an *institution* is subsidiary here to the issue of government as an *activity* involving a relationship to the self at the same time as a relationship to others. This dual relationship precisely pertains to the constitution of the subject—in other words, practices of subjectivation" (Dardot and Laval, *The New Way of the World*, 353).

13. As critic Lewis Hyde writes in relation to the eighteeenth-century enclosures, "The central meaning of enclosure's erasure of the commons lies in the way it carved those thousand-year-old beings, the commoners, into their con-

stituent parts, then reshaped them for the new world of efficiency, law, progress, and time-as-money" (Hyde, *Common as Air*, 40).

14. As a 2009 document states, "While the profit motive and ecological sustainability may conflict in the short term, in the medium and long term they form a powerfully virtuous circle, if only we can get them to work together" (Commission for Environmental Cooperation, *Green Paper on Reform of the Common Fisheries Policy*, 34).

15. The literature on the "neoliberalization of nature" rarely considers the relationship between liberalism and neoliberalism except when pointing to similarities (Heynen et al., *Neoliberal Environments*). Foucauldian analysis, particularly in terms of governmentality, is sometimes mentioned but rarely taken up as a theoretical tool to complement Marxist analysis (McCarthy and Prudham, *Neoliberal Nature and the Nature of Neoliberalism*, 280).

16. Sociologist Mariam Fraser writes: "This shift, from explaining daily experiences, as Mills describes it, to explaining the knowledges that produce experiences, does not in itself require the empirical baby to be thrown out with the experiential bathwater. On the contrary, empirical research into the conditions and effects of knowledge production—and in particular, notably, the production of the subject (who has the experience of 'having' experience but is 'in fact' constituted by it)—has proved to have considerable mileage in sociology" (Fraser, *Experiencing Sociology*, 68).

17. Castletownbere is the largest town on the Beara peninsula, County Cork, in the South West of Ireland. The town had a population of 868 in 2008, with an additional thousand people in the catchment area. It is the second-largest fishing port in Ireland and the largest whitefish port. Fishing is the main economic activity.

18. Similarly, theorist Donna Haraway criticizes the method of "science-in-action," of "following" the production of scientific knowledge and networks, as akin to miming—producing a critique that is no different from that which is critiqued, and thus being unable to go beyond it (Haraway, *Situated Knowledges*).

19. This is the objective of such methodologies. In *Reassembling the Social*, Bruno Latour suggests that the acronym ANT (Actor Network Theory) fits perfectly the role of the social researcher: "blind, myopic, workaholic, trail-sniffing, and collective traveller" (Latour, *Reassembling the Social*, 9). This corresponds with the "grounded" approach advocated within governmentality studies. As part of the same intellectual movement away from "crude structuralism," both employ a method that has been described as "topological." Instead of joining points through theoretical leaps, the job of the researcher was to travel "down into the valleys and up into the hills," tracing the movement of actors, discourses, and representations through their messy negotiations with the world.

20. In Callon's study of "translation" in a scallop fishery in Saint-Brieuc Bay, for example, power to construct reality rests with the three researchers. The role of the scallops and the fishermen is seen to be equivalent: they can either

assent or dissent. In other words the world is traced from the point of view of power (Callon, *Some Elements of a Sociology of Translation*).

21. Thus persons become, in effect, rather ill-defined constellations rattling around the world which are "not confined to particular spatio-temporal coordinates, but consist of a spread of biographical events and memories of events, and a dispersed category of material objects, traces, and leavings, which can be attributed to a person which, in aggregate, testify to agency and patienthood during a biographical career which may, indeed, prolong itself well after biological death" (Gell, *Art and Agency*, 222).

22. Anthropologist Adrian Peace makes a point that resonated with my own experience. His own ethnography of a small rural community suggested to him that despite the forces of modernity the people and place retained "a strong, indeed pervasive, sense of its own distinct identity, of being a special place in the world" (Peace, *A World of Fine Difference*, 1).

2. The End of the Line

1. This can be thought in terms of what Paul Rabinow has called a new "ethical regime of life": "The pertinent point is that the main mode of regulation is now 'ethical.' In principle, and by principle, ethical regulation operates now at the scale of living beings (*le vivant*) and takes as its task the protection of life—life and living beings that are presumed to be threatened and endangered" (*Anthropos Today*, 115). Significantly, Minister Coveney's claim that the new CFP is "real and measurable" indicates how the goal of MSY translates an ethical concern about the health of the marine environment into bioeconomic terms; achieving and maintaining the optimal productivity of fish stocks becomes a techno-managerial process once we have all agreed on the common value of marine life. This equates to what Erik Swyngedouw calls "the fantasy of 'sustainability'": the "possibility of an originally fundamentally harmonious Nature, one that is now out of synch but which, if 'properly' managed, we can and have to return to by means of a series of technological, managerial, and organisational fixes" (Swyngedouw 2010, 230).

2. I was told by a fisherman that up until the early 2000s the quotas received each year were seen as "meaningless pieces of paper" that made little difference to what fishermen in Castletownbere were catching.

3. In 2010 the Pew Environment Group commissioned a study assessing the economic, environmental, and social impacts of the decommissioning program undertaken between 2000 and 2006. The recipients (fishermen) evaluated in the study accounted for more than 90 percent of the European fisheries subsidies, amounting to 3.2 billion euros. The study found that despite promises to reduce fishing capacity in order to bring the fishing fleet into line with the available fish stocks, the effect of the financial incentives was minimal. The report concludes that overcapacity and overcapitalization of the fishing industry are the principal reasons for the failure of the CFP and that financial mea-

sures have not been effective at addressing these problems (Poseidon, *Financial Instrument for Fisheries Guidance 2000–2006*).

4. In the reign of the Tudors, for example, "forestalling" (withholding food from the hungry market in order to force the price to rise), "engrossing" (the practice of monopolizing the market for the same purpose), and, worst of all, "regrating" (buying in order to sell) were all punishable under law (Linebaugh, *The Magna Carta Manifesto*).

5. In one catch on a whitefish trawler, I noted what we kept and what we discarded: "Species kept: brill, turbot, plaice, ling, haddock, cod, monkfish, meghrim, whych, sole, lemon sole, John Dory, Pollack, skate. Species dumped: conger eel, dab, scad, everything too small, various crab species (velvet, fiddler, hermit, spider, brown), octopus, cuttlefish, pout, crayfish, red fish, wrasse, weaver fish, dogfish and many others I can't identify, as well as bits of rubbish, old net, gloves, etc." (Bresnihan 2009).

6. The reduction of discarding was first identified as an objective for the CFP in 2002. Even at this point, it was acknowledged that existing regulations (fixed quotas) made it compulsory for fishermen to discard fish that were outside their quotas at sea. This was particularly problematic in mixed fisheries where different species of fish swim together, making selective fishing very difficult. If the composition of a catch is 80 percent haddock and 20 percent cod, and there is no quota for the cod, then the cod has to be thrown overboard despite being commercially valuable and overfished. In this situation, fish that are severely depleted become by-catch and are dumped overboard. The economic incentive to discard was also identified in the 2002 CFP: "high-grading" refers to a practice whereby fish of lesser commercial value are discarded in order to make room for more commercially valuable species or sizes. While minimal efforts were made to encourage fishermen to reduce their levels of by-catch, it was not until 2007 that the European Commission issued a report recognizing that a ban on discarding was the only way to effectively reverse these regulatory and economic incentives (com 2007):

> "The problem of discards is of such importance and its repercussions of such gravity that drastic solutions must be sought now, at a time when the increasing scarcity of stocks is of great concern to Community fisheries and when the ecological damage, true or imagined, caused by fishing is raising more and more questions. The Commission acknowledges that some of the existing Community rules contain drawbacks. *The search for solutions will only be possible, however, if all the forms of and reasons for discards are considered*" (Commission for Environmental Cooperation, *Directorate-General for Research*; emphasis added).

7. Individual transferable quotas have a long and contested history in the literature on fisheries management. Ten years before the publication of Garret Hardin's well-known essay "The Tragedy of the Commons," the fisheries econ-

omist Scott Gordon argued for the necessity of property rights where a situation of open access applied. The depletion of fish resources was understood to be the inevitable outcome of unregulated access to a common resource. However, while some countries (New Zealand, Australia, Iceland) have introduced a system of individual transferable quotas, there has been significant resistance in Europe to a system that is identified as a form of enclosure or privatization. Individual transferable quotas (ITQs) are an individualized, exchangeable right of access to a resource, a right that functions as a private property claim. See Gordon, "Economic Theory of a Common-Property Resource."

8. A report commissioned by the EU in 2002, "Management of Fisheries through Systems of Transferable Rights," stated, "With increased possibilities of free transfer of commodities, people and capital throughout the EU, *international quota trade is just a matter of time*. Relative stability is an exception to principles of free movement of capital and labour" (Commission for Environmental Cooperation, *Management of Fisheries through Systems of Transferable Rights*; emphasis added).

9. A report on ITQs by EcoTrust, a Canadian NGO, echoes this pragmatic approach, blurring and softening the fact of privatization through an emphasis on its potential efficacy as part of a "holistic" form of resource management. While the report offers a critical account of the negative social and ecological consequences of ITQs in the salmon fisheries in British Columbia, it concludes that ITQs are not in themselves the problem: "Debate about ITQs is often polarized and fueled more by ideology than reality. . . . Downplayed is the critical role that sound science and good governance—that is, inclusive, transparent co-management between government, and industry and stakeholders—play in ensuring the sustainability of fisheries. The central lesson of this brief investigation into ITQ is that there are no simple solutions or quick fixes to fisheries conservation. *If properly designed, ITQ systems can play an effective role in a multi-faceted approach to responsibly managing fisheries*" (Ecotrust, *Briefing*; emphasis added).

10. Throughout the book, names of interviewees have been changed to protect anonymity.

11. Although the situation had calmed down by the time I was living in Castletownbere in 2009, the effects of this period of intense conflict continued to be felt and referred to by people I spoke to.

12. Pelagic refers to fish that live in the pelagic zone of the ocean, neither close to the bottom nor near the shore. Demersal refers to fish that live near the bottom of the ocean and thus tend to live closer to shore.

13. "At first sight, it appears that ITQ can hardly be described as real property rights. The existing systems show that ITQ can be limited in various ways. The holder cannot use them as he sees fit or dispose of them at will. In France, Parliament took a clear stance that fishing quotas cannot be considered property rights. Iceland has also stated in its fisheries management legislation that fish stocks are the property of the Icelandic nation" (Commission for Envi-

ronmental Cooperation, *Management of Fisheries through Systems of Transferable Rights*, 10).

14. This argument echoes the justifications for the Enclosure Acts of the seventeenth and eighteenth centuries. Legislation to "protect" the rivers, for example, included long prefaces that lamented the degradation of the fisheries, with the guilty party identified as those who caught fish by "destructive" means. In *The Compleat Angler*, written in the mid-seventeenth century, Izaak Walton writes: "*That which is every bodies business, is no bodies business*" (Walton, *The Compleat Angler*, 61). He is calling for restrictions to be implemented to prevent those who access the rivers for reasons other than "sport." This statement is similar to Hardin's own "Tragedy of the Commons" thesis and arguments currently being made by policy makers about the European fisheries.

15. As David Butcher of the World Wildlife Fund (WWF) stated, "We need to get over the notion that fisheries resources are the sole concern of the fishing industry. At the end of the day, fisheries are a public resource and decision-making must therefore take this public interest into account" (Mikalsen and Jentoft, *From User-Groups to Stakeholders?*).

16. Sociologist Melinda Cooper makes a similar connection between new legislative tools like the Precautionary Principle and the logic of responsibility extended through liberal rationalities. She writes, "Acting in the name of a generalized suspicion, the precautionary principle is perhaps less progressive than it might at first appear. It finds its political counterpart in neoliberal social policies that dismantle the buffers of the welfare state only to criminalize the slightest acts of deviance. Zero tolerance is the sociological face of environmental precaution" (Cooper, *Life as Surplus*, 84).

17. The EU has stated that relations between science and policy ought to reflect a "new form of governance" (Fritz, *Towards a "New Form of Governance" in Science-Policy Relations in the European Maritime Policy*). In 2002 the EU published its Science and Society Action Plan (EU). In its foreword, politician Philippe Busquin wrote, "The aim of the European Commission's Science and Society Action Plan is therefore to pool efforts at European level to develop stronger and more harmonious relations between science and society" (Busquin 2002, 3).

18. This echoes the insight of economic historian Karl Polanyi, who rightly observed the unique and circular logic contained in the appeal to the "free" market by liberal political economists: "There was nothing natural about *laissez-faire*; free markets could never have come into being merely by allowing things to take their course. To the typical utilitarian, economic liberalism was a social project which should be put into effect for the greatest happiness of the greatest number; *laissez-faire was not a method to achieve a thing, it was the thing to be achieved*" (quoted in Dardot and Laval, *The New Way of the World*, 49; emphasis added).

19. This goes some way to explaining familiar lines such as, "Strikingly, Malthusian catastrophism makes claims to be rigourously scientific and *yet*

seems impervious to falsifiability, no matter how often its claims are disproven empirically" (Yuen, "The Politics of Failure Have Failed," 29; emphasis added).

20. In 1776 Arthur Young, an agronomist and an early advocate of "improvement" and enclosure, travelled around Ireland. He gave an account of his travels in a text entitled "Tour around Ireland." Young was preoccupied with "improvements": the majority of his travelogue is spent describing either the poor conditions of the land and the way it is being (mis)used, or applauding the changes that have already been implemented by certain progressive land owners. He lists exhaustively the lack of draining projects, transport infrastructure, land management, and new agricultural techniques. He laments the impoverished condition of the majority of the rural population and ties their fate to the economic benefits (and incentives) that will arise from the "improvements" he outlines.

21. Interestingly, historian Jeanette Neeson observes that the cost of enclosure was a significant factor in determining who was able to benefit from the Enclosure Laws. The cost of fencing or ditching was prohibitive for small landholders or cottiers (see Neeson, *Commoners*).

3. Stewards of the Sea

1. The European Network for Rural Development launched LEADER in 1991 with the aim of improving the development potential of rural areas by drawing on local initiative and skills, promoting the acquisition of know-how in local integrated development, and disseminating this know-how to other rural areas. See European Network for Rural Development, LEADER Gateway, https://enrd.ec.europa.eu/en/leader.

2. As early as 1982, when the quota system was introduced, it was clear that fish catches were not going to increase indefinitely. A report presented to the Donegal County Committee of Fisheries in 1983 by the Director of the Irish Fishermen's Organisation entitled "The Marketing of Fish" opens: "I firmly believe that many of our problems could in fact be turned into opportunities. We have no choice now, but to wake up and make the best of the bad catch we have landed. The *only area capable of worthwhile expansion now is the processing sector*" (emphasis added). The 1983 report even goes on to outline similar strategies to those outlined by state policy today: finding better markets through better, more varied distribution networks; quality assurance and standards; increased consumer awareness and the development of an Irish seafood "brand."

3. BIM's report builds on two reports published by the then Department for Communications, Marine and Natural Resources in 2006: *Steering a New Course: Strategy for a Restructured, Sustainable and Profitable Irish Seafood Industry, 2007–2013*; and *Sea Change: A Marine Knowledge, Research and Innovation Strategy for Ireland, 2007–2013*. Both of these reports identified the need for a shift away from a fisheries sector based on the extraction of wild marine resources to a more strategic industry oriented toward the demands of the market. The *Cawley Report* acknowledges that achieving this vision of a "compet-

itive and market-focused seafood industry" will entail a "painful transition as the fundamental restructuring of the industry, from point of catch to onshore processing, distribution and marketing, takes shape" (Department of Communications, Marine and Natural Resources, *Steering a New Course*).

4. This is in line with EU policy more generally. The Lisbon agenda (2000) set the ambitious goal of making Europe the most competitive and dynamic knowledge-based economy in the world, capable of sustainable economic growth with more and better jobs and greater social cohesion. The European seas are understood to be a prime new territory for this development. The seas are a frontier into which the knowledge economy can expand, providing a rich source of potential. At a recent speech to the European Economic and Social Committee about the "future economics of the sea," Commissioner Damanaki said, "We know that there is clearly scope for the oceans, seas and coasts to unlock new sustainable sources of growth. And we must be capable of channelling these activities into industrial applications and ensuring that they benefit society. It is what I like to call 'Blue Growth'" (Damanaki, *The Future Economics of the Sea*).

5. The most tangible evidence of this was the investment of five million euros to set up the Seafood Development Centre in Clonakilty, County Cork. Opened in 2009, this center brings together fish producers and their raw product with market researchers and state-of-the-art facilities to produce new "ready-to-launch" convenience or ready-to-eat seafood products.

6. This "smart" and "green" approach to the agrifood business is in line with the most recent report by the Irish government on their strategy for the agrifood business, *Food Harvest 2020*. The objective is to fully capitalize on the "green" image of Ireland so that "consumers in key markets will recognize implicitly that by buying Irish they are choosing to value and respect the natural environment."

7. In 2011 Charles Glover, the author of the book on which *The End of the Line* was based, wrote a critical review of Mark Kurlansky's *The Last Fish Tale*, a sympathetic history of fishermen from below. Glover wrote that Kurlansky "remains hooked on a mythical figure in an oily sweater at a time when what makes a fisherman great is now measured by what he leaves in the sea." Glover dismisses Kurlansky's "romanticization" of fishermen, arguing that environmentalists are the ones who need to be praised. As an example, he quotes a campaign by environmentalists to close off 6,500 miles of sea to fishing, "with support from intelligent fishermen" (Glover, *Review*).

8. Seafood Services Australia (SSA) is the world pioneer of the EMS program as a means of encouraging the Australian fishing industry to embrace sustainability as an "opportunity." Their website states, "Our ability to demonstrate that we are utilizing the natural resources used by fisheries and aquaculture in a sustainable, responsible way is a cornerstone of our industry's future. . . . It is the key to our future access to these natural resources and the livelihoods of current and future generations of seafood industry operators."

9. Special Areas of Conservation (SACs) are designated by member states under the Natura 2000 framework. They belong to an EU-wide network of nature protection areas established under the 1992 Habitats Directive. The aim of the network is to assure the long-term survival of Europe's most valuable and threatened species and habitats. Under the Habitats Directive, Member States designate Special Areas of Conservation (SAC), and they also designate Special Protection Areas (SPAs) under the 1979 Birds Directive. At this time, the Irish government had failed to carry out the necessary baseline studies required for SACs and SPAs. There was pressure coming from Europe on Ireland to ensure that such areas were being sustainably managed. This led to concerns among fishermen and other resource users that they would be forced to stop their activity for a time.

10. In a later interview, another fishermen's co-op was quoted as an example. The Erris inshore fishermen's cooperative had been involved in a campaign to protect the rights of some of their members against the potential laying of a gas pipeline from an offshore site to a refinery onshore. This high-profile campaign had pitted a local community against a powerful multinational corporation. While the situation is complex, one way of framing it, as O'Sullivan did, was in terms of a conflict over resources and their value. She said the situation and its resolution offered many lessons to industry and government for the future. These lessons applied both to how government and industry should deal with communities but equally to how communities should deal with government and industry. She said the biggest lesson was the importance of gathering information and being able to communicate it. In the beginning, the Rossport community had been represented negatively, she said. When they realized they had to present themselves differently in order to reach an agreement, the information they put together was a vital "weapon." She told me that a recent Sustainable Energy Authority of Ireland (SEAI) report on offshore renewables had flagged the Erris case as a benchmark example of conflict resolution over resource use. Jean suggested to me that conflicts over resource use, as in the Corrib gas pipe example, would become more frequent and significant in the years ahead. For a more critical analysis of this "conflict resolution" process, see Garavan, "Problems in Achieving Dialogue."

11. By situating the emergence of neoliberal thought in response to the social and economic crises of the great depression, Foucault shows how classical liberal thought was insufficient for addressing these problems. The main response by governments in North America and Western Europe in the 1930s was the adoption of Keynesian policies. Broadly, these policies sought to stimulate demand through state interventions in the economy; they tried to "fix" the problems of the market by increasing wages, investing public money in large-scale infrastructure projects, redistributing wealth, and in some cases nationalizing the means of production. At the same time, however, a different school of economic thought was emerging in Germany and Austria that opposed both

classical liberal and Keynesian analysis of the crises (see Amable, *Morals and Politics in the Ideology of Neo-Liberalism*; Dardot and Laval, *The New Way of the World*). These arguments would come to influence German and North American currents of neoliberal thought in the postwar period, including responses to mounting environmental problems.

12. Even Keynesianism maintained the idea that the market was a separate sphere of activity characterized by the exchange of goods and services between buyers and sellers. "Treating the market as a natural entity really amounts to making it bear the blame for everything that does not work, playing off the 'nature' of needs against the 'nature' of the market, in short, gradually discrediting the latter in the name of the former. The State must thus intervene because of the market, in order to compensate for its deficiencies and to limit dysfunctions in the mechanism of exchange" (Donzelot, *Michel Foucault and Liberal Intelligence*, 123).

13. There is particularly interesting work going on in the area of "neoliberal conservation," which critically examines how ecosystems services, biodiversity, and the "externalities" of modern, industrial production are being accounted for through new forms of valuation and measurement. The extension of economic rationalities to this vast reserve of "natural capital" opens in turn new opportunities for commodification and financialization See Büscher et al., *Towards a Synthesized Critique of Neoliberal Biodiversity Conservation*; Sullivan, *Banking Nature?*; Robertson, *The nature that capital can see*; Turnhout et al., *Measurability in Biodiversity Governance*.

14. "While in classical liberalism there was a sustained debate over the material limits to economic growth, the neoliberal discourse is dominated by Promethean accounts of technology and economic expansion, where the case for the limits to growth is reverted into a case for the growth of limits" (Pellizzoni, *Governing through Disorder*, 796).

15. Interestingly, Malthus had already recognized that the self-interested, economic individual could not be taken for granted, as most of the liberal political economists in the nineteenth century assumed. "Hence, Malthus's interest in the *institutional preconditions* for economic success and his persistent concern with those *moral or psychological factors* which determined whether a society possessed what Hume described as 'the quick march of the spirits,' the constant spark of ignition that would consistently make riches preferable to indolence. Ricardo and his followers took such matters for granted, at least as far as Britain was concerned. *This Malthus was unwilling to do: economic man could not be assumed; he had to be nurtured*" (Winch, *Malthus*, 110; emphasis added).

16. Perhaps the clearest indicator of this simultaneous withdrawal and extension of state power comes in the figure of the "mentor" and the practice of "peer-to-peer" advice. Jean and Simon did not go into much detail about what this meant, but mentors are fishermen, or ex-fishermen, who are employed by

BIM on a casual basis to work within the industry and encourage fishermen to get involved in both the EMS project and various accreditation schemes. After our interview, Jean sent me an essay published in the *Harvard Business Review* about the "science of persuasion." She told me how the Australians went about it:

"What the Australians recommended was mentoring, industry mentoring, so it was like fisherman to fisherman so basically you don't try to reach every single fisherman because some might not like it, they might be distrustful, they'd be cautious of government, because they might not necessarily trust us. But using an industry mentor, one of their peers, you get into the whole psychological effect of peer mentoring . . . the . . . science of persuasion, and that's what it's all about, the persuasion that he's a peer, he has experience. At the moment we only have one mentor. It was on the table for a while with the coops. We said if you like why don't you all pick a champion and you can filter out that way, so someone like me would be a support to that mentor, that champion would be like an outreach, rather than me trying to track him down, he's got the flexibility then to meet them on a Saturday or Sunday, because it doesn't matter to them, its a social thing, it's a network" (O'Sullivan, interview with author).

17. In Ireland, BIM provides some funding for fisheries seeking accreditation. Co-funded by the European Fisheries Fund, the Seafood Environmental Management and Certification Grant Aid Scheme "provides a framework of support for the Irish Seafood Industry intent on achieving internationally accredited, third party audited, standards for wild capture fisheries." It provides up to a maximum of 500 euros per vessel, and from 40 to 60 percent of the implementation and certification costs, up to a maximum of 750 euros per vessel. Thus, even if a fishery is able to secure the full amount of funding, they will be required to generate considerable additional funds on their own.

18. Even when fisheries like the North West pelagic fishery have secured accreditation in the past, the ongoing conflict between Iceland and the European Union over mackerel quotas has meant that their certificate has been temporarily suspended.

19. Lockie describes how a similar governance strategy in Australia sought the "facilitation and coordination of local action" rather than the "direction or regulation of that action" (Lockie, "The State, Rural Environments, and Globalisation," 598).

20. According to sociologist Andrew Blowers, "Political modernisation and its ecological variant are, in many ways, even less democratic than the system they have supplanted. In theory, political modernisation urges the doctrine of consensus and purports to include a variety of interests. In reality it supports a system that is exclusive, elitist and unrepresentative, a condition described by Crouch (*Privatised Keynesianism*) as "post-democracy" (Blowers, *Inequality and Community and the Challenge to Modernisation*, 67). Dryzek and Niemeyer also argue that networked forms of governance are not as accountable as formal, representative politics: "Networked governance is almost impossi-

ble to render accountable in standard democratic terms because there is often no unique demos associated with a network" (Dryzek and Niemeyer, *Discursive representation*, 485).

21. When I asked Jean O'Sullivan why the EMS scheme had not been taken up more enthusiastically by fishermen, O'Sullivan explained it was because of their skepticism toward the "change in culture": "Because as far as they're [fishermen] concerned, they say what are you bothering me about, just put it into legislation and I'll . . . sure you've made everything else a legal instrument, why are you stopping now . . . like, you know, it's also a change of culture from the top down" (O'Sullivan, interview with author). In my own research, however, I found it almost impossible to identify who was responsible for different aspects of the accreditation process and where and how financial and institutional support for groups of fishermen could be secured. This was after several interviews with key individuals working in BIM on MSC accreditation.

22. This becomes particularly problematic when the more powerful actors in this new commodity chain are also unaccountable. The "Fish Fight" campaign, for example, involved a celebrity chef without much knowledge of the fisheries leading an extensive media campaign supported by other high-profile celebrities against discarding. While the campaign may have been well-intentioned, this kind of populist environmentalism can simplify what is a complex situation, as was the case with the ban on discarding. For example, on May 15, Hugh Fearnley-Whittingstall was invited to debate with Barrie Deas, head of the National Federation of Fishermen's Organization in England, on BBC *Newsnight*. He was challenged and criticized for making claims and heading media campaigns that were undermining the livelihoods of fishermen and not conveying the complexities of the fisheries and fisheries management.

23. For an account of the cultural politics involved in the construction of eco-labels and sustainable seafood and the role of celebrities in this process, see Silver and Hawkins, "'I'm Not Trying to Save Fish.'"

24. In a separate but related example, a mussel farmer I knew and worked for had gone to great lengths and expense to have his mussels audited and differentiated according to a strict set of criteria (everything from the cleanliness of the water to the quality of the mussel meat). He told me that after all that he did not get anymore money because the mussels were sold in France alongside all the others. He did not sell directly to customers, and those who bought them were only interested in the price, not the quality.

25. BIM's most recent strategy report acknowledges this deficit: "Small and medium sized seafood companies, many of which are family owned, are and will continue to be an essential component of the Irish seafood landscape. However, when we look at the scale of the market opportunity as outlined above and we consider the high level of fragmentation in our sector, it is clear that changes are needed in how our sector approaches the global seafood market. Thus, the main theme of this strategy is the need to realise scale in the sea-

food sector, to engender greater competitiveness and take advantage of the global market opportunities for seafood" (BIM, *Capturing Ireland's Share of the Global Seafood Market*).

4. Community-Managed Resources

1. Although the inshore fisheries do not fall under the auspices of the CFP, they are subject to other European environmental directives, such as the Birds and Habitats Directive and Natura 2000.

2. The Bord Iascaigh Mhara document, "Managing Access to the Irish Lobster Fishery" (2008), states that lobster stocks are overexploited in most areas around the Irish coast: "In these areas the catch per pot haul has been steadily declining for several years and, while the stock does not appear to be in imminent danger of collapse, the reduction in catch rates has resulted in fishermen using more and more pots to maintain their earnings" (BIM, *Managing Access to the Irish Lobster Fishery*, 1).

3. The first example in Ostrom's book of a successful collective response to a resource dilemma is Alanya, a Turkish inshore fishery. She goes on to discuss several other examples of community-managed inshore fisheries, including the Maine lobster fishery. It is no surprise that Ostrom should draw on the examples of locally managed fisheries to illustrate alternatives to centralized state regulation or the institutionalization of private property. Small-scale fisheries are particularly difficult to accommodate within either of these frameworks.

4. Figures from 2007 show that the inshore sector accounted for 1,400 of the 1,800 vessels in the entire fleet. The inshore sector employs about 2,300 fishermen out of a total of 4,900 in the entire sector. Despite making up such a large proportion of the fleet and fishermen, the inshore sector only lands 20 percent of the total catch-value of the Irish fisheries (Bord Iascaigh Mhara, *Managing Access to the Irish Lobster Fishery*).

5. If they are on the border of unit 8 and 9, for example, they will be able to apply for an authorization to fish in subunits 8a and 9b.

6. In 2000 Maine lobster sales were $186.1 million, with 6,884 licensed lobster fishermen operating (Acheson, *Capturing the Commons*). The closest comparable figures for Ireland are approximate sales of 13 million euros with 1,400 vessels fishing (Tully et al., *The Lobster [Homarus Gammarus L.] Fishery*).

7. Another well-known example was the collapse of the Newfoundland cod stock in the late 1980s. Despite warnings from inshore fishermen that their cod catch was declining because of the operations of larger boats offshore, the scientific data suggested otherwise. The cod stock finally collapsed and whole fishing communities lost their right to fish. These historical examples have become well-established references in reflections on science-industry relations.

8. Acheson's arguments are confirmed by the research of Fikret Berkes (1987), who has studied the fishing strategies of the Cree Indians in James Bay. He argues that even though they do not have expert management consultants, their

own practices constitute a form of ecological management over what, when, and where to fish. Interestingly, Berkes finds that Cree fishermen believe fish are inexhaustible. As a result, they use the most effective mesh size when they fish, and they fish for the most easily exploitable stocks—two measures that are anathema to current fisheries management policy. Still, they do not overexploit the stock because "they observe the proper traditions of hunting and humility toward nature."

9. "Broadly viewed," Hanna writes, "environmental problems are problems arising from incomplete and asymmetric information combined with incomplete, inconsistent, or unenforced property rights" (Hanna et al., *Rights to Nature*, 3; emphasis added).

10. This is also visible in European fisheries policy. In their 2007 report, Symes and Sissenwine wrote, "Responsibility for making things happen at the local level rests with fishing communities themselves acting independently or in a collaborative framework. *Those communities able to demonstrate initiative, leadership, the will to succeed and strong internal support through local management plans are the ones most likely to survive in the 21st century*" (Sissenwine and Symes, *Fisheries Management and Institutional Reform*; emphasis added).

11. Describing the collapse of the New England fishery in 1990, Ostrom writes, "The issue in this case—and many others—is how best to limit the use of natural resources so as to ensure their long-term economic viability" (Ostrom, *Governing the Commons*, 1).

12. Even where a "conservationist" ethos is identified, its very existence is questioned because it goes against all economic reason: "Firstly it's difficult to understand some people who are so proactive in the industry, so long term, even when they don't have a son or a daughter or whatever, they are still very conservationist. They make a living and are very active and willing. So there is a huge amount of good will there. You can't rationalize it to a certain degree, but it's just something that's there and obviously that's a very positive thing" (Carney, interview with author).

5. The More-Than-Human Commons

1. In historic terms, the activity of fishing has always been "anachronistic": it does not correspond to a clearly defined Fordist-industrial period, for example, with tightly disciplined wage-labor, or even with the period of agricultural capitalism that relied on clearly defined private rights of property. Indeed, fishermen are often compared to hunter-gatherers on the basis that their livelihoods depend on the unpredictable search for wild fish.

2. Just as European women in the seventeenth and eighteenth centuries were at the forefront of struggles for access to common lands that were being enclosed, so now in parts of the Global South women are doing the same because they are often the primary producers of food as well as the harvesters and gatherers of other resources necessary for energy and healthcare. Mies, who often

relies on stories to articulate the entwining of women, nature, and peasant within her "subsistence perspective," narrates the story of her own mother, who in the face of general shortages of food refrained from slaughtering their last pig at the end of World War II because "life must go on": she knew the pig would have piglets (Mies and Bennholdt-Thomsen, *The Subsistence Perspective*).

3. Vandana Shiva reminds us that the term *resource* "originally implied life. Its root is the Latin verb *surgere,* which evoked the image of a spring that continually rises from the ground. Like a spring, a 're-source' rises again and again, even if it has repeatedly been used and consumed" (Shiva, *Resources*, 228). While this meaning of "resource" departs from the liberal economic fixation on scarce resources, it also, she goes on, points us to the necessary human or social work involved in maintaining life-giving resources.

4. In historical terms, Linebaugh describes how evidence of the commons will often appear anecdotal, as folklore, or as "crime"—just a small story or a minor transgression that is seemingly incidental to some other, major theme. Evidence of customary commons apparently belongs to locale or craft, and therefore to trade or local histories, not "grand narratives" (Linebaugh, *The Magna Carta Manifesto*). Echoing this, Neeson writes, "If the measurement of crime has its dark figure of criminal acts not reported, common right has its dark figure too—of *practice* not recorded" (Neeson, *Commoners*, 79–80). Neeson shows us that the commons were not reducible to archaic customs or laws: this was not their only, or most significant, form. Rather, the commons were organized through everyday social practices that negotiated the varying needs, possibilities, and limits that emerged as people directly interacted with, produced, and cared for the resources they relied on.

5. This echoes John Holloway's argument that we need to develop a different sensibility for seeing what is different. He quotes John Berger: "Yet it can happen suddenly, unexpectedly, and most frequently in the half-light-of-glimpses that we catch sight of another visible order which intersects with ours and has nothing to do with it. The speed of a cinema film is 25 frames per second. God knows how many frames per second flicker past in our daily perception. But it is as if, at the brief moments I'm talking about, suddenly and disconcertingly we see between frames. We come upon a part of the visible which wasn't destined for us. Perhaps it was destined for night-birds, reindeer, ferrets, eels, whales . . ." (quoted in Holloway, *Zapatismo Urbano*, 36).

6. The notion of "territory" as developed in parts of Latin America since the early 1990s reflects something similar. In this case territory is not just an area of land or a set of resources but a historical, cultural, and socionatural entity that is based on, and sustained through, ongoing relations and interactions. This allows for "manifold logics of appropriation of the territory" by the communities who use it (quoted in Escobar, *Territories of Difference*, 58).

7. A *currach* is a small, wooden boat traditionally used all along the West Coast of Ireland.

8. The uncertainty was reinforced the second time I went out trawling, this time for whitefish. We caught decent numbers of fish for five days because of some information the skipper had heard about the fishing grounds we were on. Running out of ice, the skipper decided to land the fish, refill the hold with ice, and come back out for more fishing. By the time we got back out to the fishing grounds—a matter of six of seven hours—the weather had turned. The wind must have had an effect on the fish. In the same place we had been fishing before, we only caught tonnes of spiny redfish that had no commercial value. Not long after that, the net ripped on some rocks and there was no spare. We came back in without anything to add to the initial five days fishing.

9. In his book written in the 1930s, biologist Jakob Von Uexkull gives as an example the different ways in which an oak tree is perceived. The fox builds his lair in the roots, the owl perches in the branches, the squirrel hides in the crannies, the ant forages in the bark, the wood cutter cuts it for lumber, and the little girl is scared of it. Each confers on the tree a "functional tone." In terms of immediate experience, the tree does not appear as a "Tree" to any of them (Uexkull, *A Stroll through the World of Animals and Men*). This relativity has been taken up by thinkers like Ingold as an example of the way in which "nature" is co-produced. The emphasis on the "functional tone," however, suggests a belief in an *a priori* subject that draws meaning out of the world rather than the generation of meaning itself through the process of interaction.

10. Referring to this "shallow empiricism," philosopher Michael Halewood points out how G. H. Mead failed to grasp the radicalism of a deeply relational constructivism by falling back on the individual as the substantial unit of analysis. He quotes Mead: "The human experience with which social science occupies itself is primarily that of individuals. . . . This principle is that the individual enters into the perspective of others, in so far as he is able to take their attitudes, or occupy their points of view" (Halewood, *Introduction to Special Section on A.N. Whitehead*, 10).

11. One important and prescient figure in the development of this more-than-human, relational ontology is Alfred North Whitehead, a twentieth-century philosopher and mathematician who devoted much of his later work to exploring and critiquing the modern dualism of subject and object (Rose, *On Whitehead*; Stengers, *Experimenting with Refrains*). The presumed "bifurcation of nature," as he called it, delineated most clearly in Kantian metaphysics, understood the world to be divided into "objective" nature, the unchanging and "dumb" substrate of apprehension, and the knowing human subject who gave "form" to nature. Whitehead points to the absurdity of this dualism that, as he sarcastically writes, ultimately concludes that nature "get[s] credit which should in truth be reserved for ourselves: the rose for its scent; the nightingale for his song; and the sun for its radiance. The poets are entirely mistaken. They should address their lyrics to themselves, and turn them into odes of self-congratulation on the excellency of the human mind" (Whitehead, *Science and*

the Modern World, 54). For Whitehead, nature was neither an external world of "things" nor the result of an "internal" mediation carried out by the individual subject. Rather, *nature is always a relational achievement*, the ongoing creation of subjects and objects co-mingling and becoming more than themselves. As philosopher Philip Rose writes, "The 'things' that populate this relational world are constituted by their various relations—all beings are relational beings" (Rose, *On Whitehead*, 2).

12. Non-dualistic conceptions of nature can also be identified in descriptions of premodern European socionatural relations. Ferderici writes, "At the basis of magic was an animate conception of nature that did not admit to any separation between matter and spirit and this imagined the cosmos as a living organism, populated by occult forces, where every element was in 'sympathetic' relation with the rest" (Federici, *Caliban and the Witch*, 142).

13. While this bears a resemblance to the phenomenological perspective briefly outlined above, the difference is that subjectivity is more situated and relational, more evenly distributed between people, places, animals, and things, than the human-subject-orientated position allows.

14. Ingold makes the same point in a different context: "There is nevertheless a sense in which none of us are Westerners, and that the challenge that non-Western perspectives present to Western modes of apprehension exists at the very heart of our *own* society, in the mismatch between our shared experience of dwelling in the lived-in world and the demands placed on us by external structures of production and control that seem to leave only a residual space, divorced from culture and social life, where we can truly be ourselves" (Ingold, *Perception of the Environment Essays on Livelihood, Dwelling and Skill*, 323).

15. Capturing the experimental and uncertain dimension of this relational performance, Bruno Latour writes of how every organism "wages a bet on life" (Latour, "What Is Given in Experience?"). The sense of "waging a bet" evokes the uncertainty of acting, an uncertainty that does not derive from an individual sense of doubt or ignorance but from the inherent unknowability of a world that is always experienced in relation to other people, animals, plants, and artifacts. Thus, while people may act in order to achieve a particular aim or goal, the reality that unfolds is rarely what was imagined or intended. More generally, much activity is object-less in the sense that it is not oriented by anything more than the present and the limits and possibilities that constantly arise out of the present. There is not a thinking subject, an "I," constructing the world, identifying it from a position of separation. Actions and decisions are taken without knowing exactly where they will lead. As philosopher Judith Butler writes, "It is as if every living being has tentacles of one kind or another and these prehensive activities are part of the organism, not something added on and certainly not something inessential. They are aspects of the organism, they are also modes of relationality; open-ended, not fully knowing, bound to the world, seeking to further existence on yet one more occasion" (Butler, "On This Occasion," 9).

16. Neeson offers a poetic description of this commoning sensibility and the ways it falls outside the economic (and aesthetic) gaze of "improvement" in the eighteenth century: "Sauntering after a grazing cow, snaring rabbits and birds, fishing, looking for wood, watercress, nuts or spring flowers, gathering teazles, rushes, mushrooms or berries, and cutting peat and turves were all part of a commoning economy and a commoning way of life invisible to outsider" (Neeson, *Commoners*, 40). "Sauntering" nicely captures the movement of the commoner as s/he treads a nonlinear path through the landscape, always open to the intrusion of others.

17. Ingold uses the term "taskscape" to describe the way places and people are co-produced through activities. He uses Pieter Bruegel's *The Harvesters* (1565) to illustrate this. Twelve paintings depict the months of the year. He describes how the place and the people are bound up with each other through countless different activities. He writes, "It does not begin here (with a preconceived image) and end there (with a finished artifact), but is *continuously going on*" (Ingold, *Perception of the Environment Essays on Livelihood, Dwelling and Skill*, 205).

18. A direct parallel can be made between Neeson's observation about claims for compensation in the eighteenth century at the time of enclosure, and claims for authorizations to fish lobsters in the Irish inshore fisheries. Just as lobster fishermen must prove they have a track record of fishing for lobster, Neeson writes, communities that had lost access to common lands only received compensation if they provided material proof or evidence that they had a customary right. This evidence was often created to meet the legal requirements, when in fact the right of access was enshrined in the more complex pattern of customary practices that were *performed*, not written on paper.

19. I was told by one older fishermen, for example, how each summer season he would go out with his father lobster potting and they would work their way up the bay targeting one square of sea at a time, taking note of the best fishing grounds.

20. Ostrom quotes a similar example of how fishing is policed in Davis. Here, a fisherman sees another setting his gear near or even over his own gear. He calls the other fisherman across the radio and complains. Someone else on the radio agrees and the "transgressor" puts his gear somewhere else (Ostrom, *Governing the Commons*).

21. Neeson and other social historians have described something similar when referring to the existence of "customary rights" governing the use of land, rivers, turf-moors, and forest. These were the rights of pasturage, piscary, turbary, and estovers. "Custom" is another word for practice or usage, and such rights were fleshed out through elaborate practices of "stinting." Stinting was not just about access but also about use, about recognizing and acting in accordance with particular limits. These limits were not "absolute," but were situated, reflecting the social and ecological knowledge that was required to take

care of the commons: for example, when to harvest, how to harvest, who would harvest, and how much to take. Stinting was a practical, social regime for sharing the product of the commons that recognized the needs and participation of humans and nonhumans in the production process.

22. He is referring here to the Andean concept of *Uyway*: "Embedded in everyday practices, *uyway* refers to mutual relations of care among humans and also with other-than-human beings" (De la Cadeña, "Indigenous Cosmopolitics," 354).

23. Starhawk encapsulates this move from abundance and diversity to scarcity and homogeneity when she writes, "Enclosed land, instead of serving multiple needs and purposes, served only one. When a forest was cut down and enclosed for grazing land, it no longer provided wood for fuel and building, acorns for pigs, a habitat for wild game, a source of healing herbs, or shelter for those who were driven to live outside the confines of town and village. When a fen was drained to provide farmland, it no longer provided a resting place or nesting sites for migratory birds, or a source of fish for the poor" (Starhawk, "Appendix A," 189).

6. Conclusion

1. Mayhew points out that Malthus considered the second edition of *Essay on Population* to be an entirely new book because it was filled with new empirical data and research, partly assembled from his travels in Scandinavia. He also shows that Malthus's arguments had changed: Malthus vacillated between opposition and support for the Poor Laws, and he was a prominent advocate of the Corn Laws because they ensured food security for the people of Britain.

2. For this reason we have to be critical of positions that set up an easy binary between Capitalism and Nature (see Klein, *This Changes Everything*). The capacity for ecological crises and contradictions to be "fixed" through new configurations of the state and the market is what needs to be better understood and contested.

3. "Western discourses regarding the relation to nature have frequently swung on a pendulum between cornucopian optimism and triumphalism on one pole and unrelieved pessimism not only of our powers to escape from the clutches of naturally imposed limits but even to be autonomous beings outside of nature-driven necessities at the other pole" (Harvey, *Justice, Nature and the Geography of Difference*, 149).

4. In their formulation of the concept of "liberation ecology" as a way beyond Eurocentric, capitalist systems of development, Peet and Watts write: "Hence an opening of poststructuralist materialist thought to the world of environmental experiences is as much an exercise in *critique* as it is an appeal to the virtues of local or subaltern knowledges" (Peet and Watts, *Liberation Ecologies*, 262).

5. Paul Rabinow describes this as a "practice of making": "This sensibility takes the mode of a keen awareness that the taken-for-granted can change, that

new entities appear, that our practices of making are closely linked to those entities, that we name them, that we group them, that we experiment with them, that we discover different contours when deploying questions and techniques. When this sensibility becomes reflexive, it becomes an aesthetic not of taste nor of beautiful appearance but of the invention of new sensibilities, new concepts, new techniques and ideas of techniques in response to those incommensurabilities that question our practices and that eventualize our relation to them" (Rabinow, *Anthropos Today*, 67).

6. "The only question that is really worth posing is whether the Left can counter pose an alternative governmentality to neo-liberal governmentality. At the end of his lecture of 31 January on *The Birth of Biopolitics*, Foucault asked if something like an 'autonomous socialist governmentality' had ever existed. His answer was unambiguous: such a form of governmentality has always been wanting. . . . What Foucault claims is that it is *not to be found* within socialism and its texts. And since it cannot be found, 'it must be invented' (Dardot and Laval, *The New Way of the World*, 347).

7. This echoes Nicholas Blomley's analysis of an anti-enclosure struggle in Vancouver (Blomley, *Making Private Property*). While the urban architecture slated for commercial development was considered "abandoned," it was layered over with years of historical memory and experience, experiences that continued to materialize through the everyday lives of those who lived near the building. The campaign to resist the "enclosure" of the building enacted these qualities by using and reinhabiting the space. This was not just an identifiable "community" fighting for an urban "resource," but rather a material politics that demonstrated and referred to the manifold relations and values that were constituted over time through the use of the space. Trying to find some way of expressing this common right or value, Blomley settles on the "right to not-be-excluded"—a negative right of the commons for those who cannot adequately represent its concerns and experiences within existing categories (of ownership) but nonetheless can still resist the claim of any individual or collective to exclude others.

8. The struggle for various ecological commons, many different socionatures, requires different ways of articulating our relationships with each other and nonhuman nature. As John Holloway writes in the context of the Zapatistas, a struggle for the commons has placed poetic expression at the heart of its politics: "What we know is that the realism of power politics failed to achieve radical social change and that hope lies in breaking reality, in establishing our own reality, our own logic, our own language, our own colours, our own music, our own time, our own space. That is the core of the struggle not only against 'them' but against ourselves, that is the core of the Zapatista resonance" (Holloway, *Zapatismo Urbano*, 177). Similarly, De la Cadeña adopts the idea of "slowing down reasoning" from Isabelle Stengers: "Emerging through a deep, expansive, and simultaneous crisis of colonialism and neoliberalism—converging in

its ecological, economic, and political fronts—the public presence of unusual actors in politics is at least thought provoking. It may represent an epistemic occasion to 'slow down reasoning'. . . and, rather than asserting, adopt an intellectual attitude that proposes and thus creates possibilities for new interpretations" (De la Cadeña, *Indigenous Cosmopolitics in the Andes*, 336).

9. As Geographer Erik Swyngedouw writes, "To the extent that the current post-political condition, which combines apocalyptic environmental visions with a hegemonic neoliberal view of social ordering, constitutes one particular fiction (one that in fact forecloses dissent, conflict, and the possibility of a different future), there is an urgent need for different stories and fictions that can be mobilized for realization. This requires foregrounding and naming different socio-environmental futures, making the new and impossible enter the realm of politics and of democracy, and recognizing conflict, difference, and struggle over the naming and trajectories of these futures" (2007, 36). Arturo Escobar also points to the importance of storytelling and narrative for a politics of the commons: "Storytelling and analysis must be generated around the commons in order to replace the language of efficiency with that of sufficiency, the cultural visibility of the individual with that of community" (Escobar, *Encountering Development*, 198).

10. Supporting this argument, historian Robert Marzac writes of eighteenth-century enclosure and resistance: "The land was re-presented as a raw material in need of 'improvement' and 'cultivation,' *and no longer as an entity that gave life to 'inhabitants,' a key term used by those who resisted enclosures*" (Marzac, *Energy Security*, 83; emphasis added).

11. Echoing the importance of experiment and construction, rather than more defensive postures, theorist Timothy Morton writes: "We destroy atavism and religious sincerity, in caring for the environment in the name of delight and passion, constantly making room for more writing, more songs and more care for our world" (Morton, *Ecologocentrism*, 55).

Bibliography

Aalbers, Manuel B. 2013. "Neoliberalism Is Dead . . . Long Live Neoliberalism!" *International Journal of Urban and Regional Research* 37 (3): 1083–90.

Acheson, James M. 2003. *Capturing the Commons: Devising Institutions to Manage the Maine Lobster Industry*. Hanover NH: University Press of New England.

Acheson, James M., and Bonnie J. McCay, eds. 1990. *The Question of the Commons. The Culture and Ecology of Communal Resources*. Tucson: University of Arizona Press.

Agrawal, Arun. 2001. "Common Property Institutions and Sustainable Governance of Resources." *World Development* 29 (10): 1649–72

———. 2002. "Indigenous Knowledge and the Politics of Classification." *International Social Science Journal* 54: 287–97.

———. 2003. "Sustainable Governance of Common-Pool Resources: Context, Methods, and Politics." *Annual Review of Anthropology* 32: 243–62.

———. 2006. *Environmentality: Technologies of Government and the Making of Subjects*. Durham: Duke University Press.

Ainley, David, Paul Dayton, Sidney Holt, Jeremy Jackson, Jennifer Jacquet, and Daniel Pauly. 2010. "Seafood Stewardship in Crisis." *Nature* 467: 28–29.

Amable, Bruno. 2011. "Morals and Politics in the Ideology of Neo-Liberalism." *Socio-Economic Review* 9 (1): 3–30.

Amit, Vered, ed. 2003. *Constructing the Field: Ethnographic Fieldwork in the Contemporary World*. New York: Routledge.

Auer, Mathew R. 2014. "Collective Action and the Evolution of Social Norms: The Principled Optimism of Elinor Ostrom." *Journal of Natural Resources Policy Research* 6 (4): 265–71.

Bakker, Karen J. 2003. "A Political Ecology of Water Privatization." *Studies in Political Economy* 70: 35–58.

———. 2005. "Neoliberalizing Nature? Market Environmentalism in Water Supply in England and Wales." *Annals of the Association of American Geographers* 95 (3): 542–65.

———. 2008. "The Ambiguity of Community: Debating Alternatives to Private-Sector Provision of Urban Water Supply." *Water Alternatives* 1 (2): 236–52.

———. 2010. "The Limits of 'Neoliberal Natures': Debating Green Neoliberalism." *Progress in Human Geography* 34 (6): 715–35.

Baland, Jean-Marie, and Jean-Philippe Platteau. 1996. *Halting Degradation of Natural Resources: Is There a Role for Rural Communities?* Rome: Food and Agriculture Organization of the United Nations.

Barad, Karen. 1999. "Agential Realism." In *The Science Studies Reader*, edited by Mario Biagioli, 1–11. New York: Routledge.

———. 2004. "Posthumanist Performativity: Toward an Understanding of How Matter Comes to Matter." *Signs* 28 (3): 801–31.

Barca, Stefania. 2007. "Enclosing the River: Industrialisation and the 'Property Rights' Discourse in the Liri Valley (South of Italy), 1806–1916." *Environment and History*, 3–23.

Barrell, John. 2010. *The Idea of Landscape and the Sense of Place 1730–1840. An Approach to the Poetry of John Clare*. Cambridge, UK: Cambridge University Press.

Barry, Andrew. 2001. *Political Machines: Governing a Technological Society*. London: Athlone Press.

Bate, Jonathan. 2003. *John Clare. A Biography*. London: Picador.

BBC News. 2013, May 31. http://www.bbc.com/news/world-europe-22717796.

Beck, Ulrich. 1992. *Risk Society: Towards a New Modernity*. Cambridge: Polity.

———. 2010. "Climate Change, or How to Create a Green Modernity?" *Theory, Culture and Society* 27 (2–3): 254–66.

Bennett, Jane. 2010. *Vibrant Matter: A Political Ecology of Things*. Durham and London: Duke University Press.

Berkes, Fikret. 1987. "Common Property Resource Management and Cree Indian Fisheries in Subarctic Canada." In *The Question of the Commons: The Culture and Ecology of Communal Resources*, edited by James M. Acheson and Bonnie J. McCay, 66–91. Tucson: University of Arizona Press.

Berkes, Fikret, and Carl Folke, eds. 1998. *Linking Social and Ecological Systems. Management Practices and Social Mechanisms for Building Resilience*. Cambridge, UK: Cambridge University Press.

Berkes, Fikret, and Robert S. Pomeroy. 1997. "Two to Tango: The Role of Government in Fisheries Management." *Marine Policy* 21 (5): 465–80.

Berkes, Fikret, Johan Colding, and Carl Folke. 2000. "Rediscovery of Traditional Ecological Knowledge as Adaptive Management." *Ecological Applications* 10 (5): 1251–62.

Blackman, Lisa, John Cromby, Derek Hook, Dimitris Papadopoulos, and Valerie Walkerdine. 2008. "Creating Subjectivities." *Subjectivity* 22 (1): 1–27.

Blaikie, Piers. 1985. *The Political Economy of Soil Erosion in Developing Countries*. London: Longman.

Blomley, Nicholas. 2007. "Making Private Property: Enclosure, Common Right and the Work of Hedges." *Rural History* 18 (1): 1–21.

———. 2008. "Enclosure, Common Right and the Property of the Poor." *Social & Legal Studies* 17: 311–31.

Blowers, Andrew. 2003. "Inequality and Community and the Challenge to Modernisation: Evidence from the Nuclear Oases." In *Just Sustainabilities: Development in an Unequal World*, edited by Julian Agyeman, Robert D. Bullard, and Bob Evans, 64–80. London: Earthscan.

Boal, Iain. 2006. "Specters of Malthus: Scarcity, Poverty, Apocalypse." Interview by David Martinez. http://www.counterpunch.org/2007/09/11/specters-of-malthus-scarcity-poverty-apocalypse/.

Bond, Patrick. 2012. "Emissions Trading, New Enclosures and Eco-Social Contestation." *Antipode* 44 (3): 684–701.

Bord Iascaigh Mhara. 1999. *Irish Inshore Fisheries Sector. Review and Recommendations*. Dublin: Bord Iascaigh Mhara.

———. 2005. *The Management Framework for Shellfisheries in Ireland*. Dublin: Bord Iascaigh Mhara.

———. 2008. *Managing Access to the Irish Lobster Fishery*. Dublin: Bord Iascaigh Mhara.

———. 2013. *Capturing Ireland's Share of the Global Seafood Market*. Dublin: Bord Iascaigh Mhara.

Brenner, Neil, Jamie Peck, and Nik Theodore. 2010. "After Neoliberalization?" *Globalizations* 7 (3): 327–45.

Brenner, Neil, and Nik Theodore. 2002. "Cities and the Geographies of 'actually Existing Neoliberalism'." *Antipode* 34 (3): 349–79.

Bresnihan, Patrick. 2009a. Field notes, September 15.

———. 2009b. Field notes, December 15.

———. 2013. "John Clare and the Manifold Commons." *Environmental Humanities* 3: 71–91.

Bresnihan, Patrick, and Michael Byrne. 2015. "Escape into the City: Everyday Practices of Commoning and the Production of Urban Space in Dublin." *Antipode* 47 (1): 36–54.

Brockington, Dan, and Rosaleen Duffy. 2010. "Capitalism and Conservation: The Production and Reproduction of Biodiversity Conservation." *Antipode* 42 (3): 469–84.

Brockling, Ulrich, Susanne Krasman, and Thomas Lemke, eds. 2010. *Governmentality: Current Challenges and Future Issues*. London: Routledge.

Brody, Hugh. 2002. *The Other Side of Eden: Hunters, Farmers, and the Shaping of the World*. New York: MacMillan.

Browne, R. M., J. P. Mercer, and M. J. Duncan. 2001. "An Historical Overview of the Republic of Ireland's Lobster (Homarus Gammarus Linnaeus) Fishery, with Reference to European and North American (Homarus America-

nus Milne Edwards) Lobster Landings." In *Coastal Shellfish—A Sustainable Resource*, 49–62. Netherlands: Springer.
Budds, Jessica. 2004. "Power, Nature and Neoliberalism: The Political Ecology of Water in Chile." *Singapore Journal of Tropical Geography* 25: 322–42.
Brundtland, Gru, Mansour Khalid, Susanna Agnelli, Sali Al-Athel, Bernard Chidzero, Lamina Fadika, Volker Hauff et al. 1987. *Our Common Future*. Oxford: United Nations World Commission on Environment and Development.
Burchell, Gary. 1996. "Liberal Government and Techniques of the Self." In *Foucault and Political Reason*, edited by Andrew Barry, Thomas Osborne, and Nikolas Rose, 19–36. London: UCL Press.
Büscher, Bram, Sian Sullivan, Katja Neves, Jim Igoe, and Dan Brockington. 2012. "Towards a Synthesized Critique of Neoliberal Biodiversity Conservation." *Capitalism Nature Socialism* 23 (2): 4–30.
Busquin, Philippe. 2002. Foreword. *Science and Society Action Plan*. Luxembourg: Office for Official Publications of the European Communities.
Butler, Judith. 1999. "On This Occasion." In *Butler on Whitehead: On the Occasion*, edited by Roland Faber, Michael Halewood, and Deena Lin, 3–17. Lanham MD: Rowman and Littlefield.
Caffentzis, George. 2004. "A Tale of Two Conferences: Globalization, the Crisis of Neoliberalism and Question of the Commons." Center for Global Justice. http://www.globaljusticecenter.org/commons.
———. 2008. "Autonomous Universities and the Making of the Knowledge Commons." Russell Scholar Lecture IV, November 18.
———. 2010. "The Future of 'The Commons': Neoliberalism's 'Plan B'or the Original Disaccumulation of Capital?" *New Formations* 69 (1): 23–41.
Callon, Michel. 1986. "Some Elements of a Sociology of Translation: Domestication of the Scallops and the Fishermen of St. Brieuc Bay." In *Power, Action and Belief: A New Sociology of Knowledge?*, edited by John Law, 193–223. London: Routledge.
———. 2007. "An Essay on the Growing Contribution of Economic Markets to the Proliferation of the Social." *Theory, Culture and Society* 24: 139–63.
Campling, Liam, Elizabeth Havice, and Penny McCall Howard. 2012. "The Political Economy and Ecology of Capture Fisheries: Market Dynamics, Resource Access and Relations of Exploitation and Resistance." *Journal of Agrarian Change* 12 (2–3): 177–203.
Carney, James. 2009. Interview with author, August 12. Galway.
Casey, Simon. 2009. Interview with author, June 30. Dublin.
Castree, Noel. 2008a. "Neoliberalising Nature: Processes, Effects, and Evaluations." *Environment and Planning A* 40 (1): 153–73.
———. 2008b. "Neoliberalising Nature: The Logics of Deregulation and Reregulation." *Environment and Planning A* 40 (1): 131–52.
Chazkel, Amy, and David Serlin. 2011. "New Approaches to Enclosures: Introduction." *Radical History Review* 109: 1–12.

Clare, John. 2003. *Selected Poems*. Edited by Jonathan Bate. London: Faber and Faber.

———. 2004. *Major Works*. Edited by Eric Robinson and David Powell. New York: Oxford University Press.

Clausen, Rebecca, and Brett Clark. 2005. "The Metabolic Rift and Marine Ecology: An Analysis of the Ocean Crisis within Capitalist Production." *Organization & Environment* 18 (4): 422–44.

Coelho, Manuel P., Jose A. Filipe, and Manuel A. Ferreira. 2011. "Rights Based Management and the Reform of the Common Fisheries Policy: The Debate." *International Journal of Latest Trends in Finance & Economic Sciences* 1 (1): 16–22.

Collier, Stephen J., and Aihwa Ong, eds. 2005. *Global Assemblages: Technology, Politics, and Ethics as Anthropological Problem*. Oxford: Blackwell.

Collins, Harry M., and Robert Evans. 2002. "The Third Wave of Science Studies: Studies of Expertise and Experience." *Social Studies of Science* 32 (2): 235–96.

Commission for Environmental Cooperation. 1992. *Report from the Commission to the Council on the Discarding of Fish in Community Fisheries: Causes, Impact, Solutions*. Sec. (92)423 final. European Commission.

———. 2002. *Directorate-General for Research. Management of Fisheries through Systems of Transferable Rights*. Working Paper FISH III EN. European Commission.

———. 2007. *Communication from the Commission on rights-based management tools in fisheries*. COM 173 final. European Commission.

———. 2008a. *A European Strategy for Marine and Maritime Research: A Coherent European Research Area Framework in Support of a Sustainable Use of Oceans and Seas*. European Commission.

———. 2008b. *Common Fisheries Policy: A User's Guide*. Luxembourg: Office for Official Publications of the European Communities.

———. 2009. *Green Paper on Reform of the Common Fisheries Policy*. COM 163 final. European Commission.

———. 2011a. *CFP Reform-the Discard Ban*. European Commission.

———. 2011b. *Fisheries: The EU "Zero Tolerance" Campaign against Illegal Fishing Gets Tougher*. IP/11/465. European Commission.

———. 2011c. *Impact Assessment of Discard Reducing Policies* DRAFT FINAL *Report. Studies in the Field of the Common Fisheries Policy and Maritime Affairs*. Lot 4: Impact Assessment Studies Related to the CFP.

———. 2013. "Maritime Affairs and Fisheries." *Reform of the Common Fisheries Policy: A Sustainable Future for Fish and Fishermen*. http://ec.europa.eu/dgs/maritimeaffairs_fisheries/magazine/en/policy/reform-common-fisheries-policy-sustainable-future-fish-and-fishermen.

Connelly, Michael F., and Jean D. Clandinin. 2000. *Narrative Inquiry Experience and Story in Qualitative Research*. San Francisco: Jossey-Bass.

Coole, Diana, and Samantha Frost. 2010. "Introduction." In *New Materialisms: Ontology, Agency and Politics*, edited by Diana Coole and Samantha Frost. Durham NC: Duke University Press.

Cooper, Melinda. 2008. *Life as Surplus: Biotechnology and Capitalism in the Neoliberal Era*. University of Washington Press.

Coveney, Simon. 2011. *Fisheries Council-Press Conference*. http://video.consilium.europa.eu/webcast.aspx?ticket=775-979-12874.

Crean, Kevin, and David Symes. 1994. "The Discards Problem: Towards a European Solution." *Marine Policy* 18 (5): 422–34.

———. 1995. "Privatisation of the Commons: The Introduction of Individual Transferable Quotas in Developed Fisheries." *Geoforum* 26 (2): 175–85.

Crean, Kevin, and Steve J. Wisher. 2000. "Is There the Will to Manage Fisheries at a Local Level in the European Union? A Case Study from Shetland." *Marine Policy* 24 (6): 471–81.

Crouch, Colin. 2009. "Privatised Keynesianism: An unacknowledged policy regime." *The British Journal of Politics & International Relations* 11 (3): 382–399.

Crutzen, Paul J., and Eugene F. Stoermer. 2000. "The Anthropocene." *Global Change* 41: 17–18.

Dale, Gareth. 2012. "Adam Smith's Green Thumb and Malthus's Three Horsemen: Cautionary Tales from Classical Political Economy." *Journal of Economic Issues* 46 (4): 859–80.

Damanaki, Maria. 2010. *Priorities for Maritime Affairs and Fisheries*. SPEECH/10/241.

———. 2011a. *Discarding: Key Challenge in Fisheries Policy Reform*. SPEECH/11/136.

———. 2011b. *The Future Economics of the Sea. Meeting of the Employers' Group of the European Economic and Social Committee Brussels*. SPEECH/11/178.

———. 2011c. *The New Common Fisheries Policy: Making Things Simpler*. SPEECH/11/191.

———. 2011d. *The Future of the Common Fisheries Policy Is Now*. SPEECH/11/418.

———. 2011e. *Save the Fish to Save the Fishermen*. Open editorial, CFP Reform Watch.

Dardot, Pierre, and Christian Laval. 2013. *The New Way of the World: On Neoliberal Society*. London: Verso.

Davis, Anthony. 1984. "Property Rights and Access Management in the Small-Boat Fishery: A Case Study from Southwest Nova Scotia." In *Atlantic Fisheries and Coastal Communities: Fisheries Decision-Making Case Studies*, edited by Cynthia Lamson and Arthur J. Hanson, 133–64. Halifax; Dalhousie: Ocean Studies Programme.

De Angelis, Massimo. 2001. "Marx and Primitive Accumulation: The Continuous Character of Capital's Enclosures." *The Commoner* 2 (1): 1–22.

———. 2007. *The Beginning of History: Value Struggles and Global Capital*. London: Pluto Press.

——. 2013. "Does Capital Need a Commons Fix?" *Ephemera: Theory & Politics in Organization* 13 (3): 603–15.

De Angelis, Massimo, and David Harvie. 2014. "The Commons." In *The Routledge Companion to Alternative Organization*, edited by Martin Parker, George Cheney, Valerie Fournier, and Chris Land, 280–94. London: Routledge.

De Courcy Ireland, John. 1981. *Ireland's Sea Fisheries: A History*. Dublin: The Glendale Press.

De la Cadeña, Marisol. 2010. "Indigenous Cosmopolitics in the Andes: Conceptual Reflections beyond 'Politics.'" *Cultural Anthropology* 25 (2): 334–70.

Department of Communications, Marine and Natural Resources. 2006. *The Cawley Report, Steering a New Course for the Irish Seafood Industry, 2007–2013*.

Dietz, Thomas, Elinor Ostrom, and Paul C. Stern. 2003. "The Struggle to Govern the Commons." *Science* 302 (5652): 1907–12.

Dillon, John. 1955. Dail Eireann debates, July 14, vol. 152. National Library of Ireland, Dublin.

Donzelot, Jacques. 2008. "Michel Foucault and Liberal Intelligence." *Economy and Society* 37 (1): 115–34.

Donovan, Jill. 2010. Interview with author, January 26. Cork.

Drummond, Ian, and Terry Marsden. 1995. "Regulating Sustainable Development." *Global Environmental Change* 5: 51–63.

Dryzek, John S., and Simon Niemeyer. 2008. "Discursive Representation." *American Political Science Review* 102 (04): 481–93.

Economist. 2011. "The Anthropocene: A Man-Made World." http://www.economist.com/node/18741749.

EcoTrust. 2009. *Briefing: A Cautionary Tale about ITQs in BC Fisheries*. http://www.gmri.org/upload/files/9_Cautionary_Tale_ITQs.pdf.

Eden, Sally. 2009. "The Work of Environmental Governance Networks: Traceability, Credibility and Certification by the Forest Stewardship Council." *Geoforum* 40 (3): 383–94.

Ehrlich, Paul R. 1968. *The Population Bomb*. New York: Sierra Club/Ballantine Books.

Escobar, Arturo. 1995. *Encountering Development: The Making and Unmaking of the Third World*. Princeton, NJ: Princeton University Press.

——. 1999. "After Nature: Steps to an Antiessentialist Political Ecology 1." *Current Anthropology* 40 (1): 1–30.

——. 2008. *Territories of Difference: Place, Movements, Life, Redes*. Durham NC: Duke University Press.

Esteva, Gustavo. 2014. "Commoning in the New Society." *Community Development Journal* 49 (S1): 144–59.

Federici, Silvia. 2001. "Feminism and the Politics of the Commons." *The Commoner*. http://andandand.org/pdf/federici_feminism_politics_commons.pdf.

——. 2004. *Caliban and the Witch*. Brooklyn NY: Autonomedia.
——. 2012. *Revolution at Point Zero: Housework, Reproduction, and Feminist Struggle*. Oakland CA: PM Press.
Ferguson, James. 2010. "The Uses of Neoliberalism." *Antipode* 41 (S1): 166–84.
Fine, Ben. 2009. "Financialization and Social Policy." Presented at the UNRISD conference on the "Social and Political Dimensions of the Global Crisis: Implications for Developing Countries," Geneva. http://eprints.soas.ac.uk/7984/1/unrisdsocpol.pdf.
Fisher, Mark. 2009. *Capitalist Realism: Is There No Alternative?*. London: Zero Books.
Foley, Paul, and Karen Hébert. 2013. "Alternative Regimes of Transnational Environmental Certification: Governance, Marketization, and Place in Alaska's Salmon Fisheries." *Environment and Planning A* 45 (11): 2734–51.
Food and Agriculture Organization. 2010. *The State of World Fisheries and Aquaculture*. Rome: Food and Agriculture Organization of the United Nations.
Forsyth, Tim. 2003. *Critical Political Ecology: The Politics of Environmental Science*. Routledge, London.
Forsyth, Tim, and Craig Johnson. 2014. "Elinor Ostrom's Legacy: Governing the Commons, and the Rational Choice Controversy." *Development and Change* 45 (5): 1093–1110.
Foucault, Michel. 1980. *Power/Knowledge Selected Interview and Other Writings 1972–1977*. Brighton: Harvester Press.
——. 1991. "Governmentality." In *The Foucault Effect: Studies in Governmentality*, edited by Gary Burchell, Colin Gordon, and Peter Miller, 87–104. London: Harvester Wheatsheaf.
——. 1998. *The History of Sexuality, Vol. 1: The Will to Knowledge*. London: Penguin Books.
——. 2004. *Security, Territory, Population. Lectures at the College de France, 1977–1978*. New York: Palgrave MacMillan.
——. 2008. *The Birth of Biopolitics. Lectures at the College de France, 1978–1979*. New York: Palgrave MacMillan.
Francis, Robert A., and Michael K. Goodman. 2009. "Post-Normal Science and the Art of Nature Conservation." *Journal for Nature Conservation* 18 (2): 89–105.
Fraser, Mariam. 2009. "Experiencing Sociology." *European Journal of Social Theory* 12 (1): 63–81.
Freire, Juan, and Antonio García-Allut. 2000. "Socioeconomic and Biological Causes of Management Failures in European Artisanal Fisheries: The Case of Galicia (NW Spain)." *Marine Policy* 24 (5): 375–84.
Fritz, Jan-Stefan. 2010. "Towards a 'New Form of Governance' in Science-Policy Relations in the European Maritime Policy." *Marine Policy* 34 (1): 1–6.

Funtowicz, Silvio, and Jerome Ravetz. 1993. "Science for the Post-Normal Age." *Futures* 25 (7): 739–55.
Furlong, Brian. 2012. "€10K of Fish Given Away in Protest over Quotas." *The Irish Examiner*, October 5.
Garavan, Mark. 2008. "Problems in Achieving Dialogue: Cultural Misunderstandings in the Corrib Gas Dispute." In *Environmental Argument and Cultural Difference. Locations, Fractures and Deliberations*, edited by Ricca Edmondson and Henrike Rau. Bern: Peter Lang.
Gell, Alfred. 1998. *Art and Agency*. Oxford: Clarendon.
Gibson-Graham, Julie Katherine. 2003. "An Ethics of the Local." *Rethinking Marxism* 15 (1): 49–74.
———. 2006. *A Post-Capitalist Politics*. Minneapolis: University of Minnesota Press.
Gibson, James J. 1950. *The Perception of the Visual World*. Boston: Houghton Mifflin Company.
Glover, Charles. 2011. Review of *The Last Fish Tale* by Mark Kurlansky. *The Daily Telegraph*, May 12.
Goldman, Michael. 1997. "'Customs in Common': The Epistemic World of the Commons Scholars." *Theory and Society* 26 (1): 1–37.
———. 1998. "Inventing the Commons: Theories and Practices of the Commons' Professional." *Privatizing Nature. Political Struggles for the Global Commons*, 20–53.
———. 2004. "Imperial Science, Imperial Nature: Environmental Knowledge for the World (Bank)." In *Earthly Politics. Local and Global in Environmental Governance*, edited by Sheila Jasanoff and Marybeth L. Martello. Cambridge MA: MIT Press.
Goldstein, Jesse. 2013. "*Terra Economica*, Waste and the Production of Enclosed Nature." *Antipode* 45 (2): 357–75.
Gordon, Scott H. 1954. "The Economic Theory of a Common-Property Resource: The Fishery." *The Journal of Political Economy*, 124–42.
Graeber, David. 2011. *Debt: The First 5,000 Years*. Brooklyn NY: Melville House.
Graham, Michael. 1943. *The Fish Gate*. London: Faber.
Gulbrandsen, Lars H. "The Emergence and Effectiveness of the Marine Stewardship Council." *Marine Policy* 33: 664-60.
Gupta, Akhil, and James Ferguson. 1997. "Discipline and Practice: 'The Field' as Site, Method, and Location in Anthropology." In *Anthropological Locations: Boundaries and Grounds of a Field Science*, edited by Akhil Gupta and James Ferguson, 1–46. University of California Press.
Guthman, Julie. 2007. "The Polanyian Way? Voluntary Food Labels as Neoliberal Governance." *Antipode* 39 (3): 456–78.
———. 2008. "Neoliberalism and the Making of Food Politics in California." *Geoforum* 39 (3): 1171–83.

Hajer, Maarten. 1995. *The Politics of Environmental Discourse: Ecological Modernization and the Policy Process*. Oxford: Clarendon Press.

———. 1996. "Ecological Modernisation as Cultural Politics." In *Risk, Environment and Modernity: Towards a New Ecology*, edited by Scott Lash, Bronislaw Szerszynski, and Brian Wynne, 246–68. London: Sage.

———. 2003. "Policy without Polity? Policy Analysis and the Institutional Void." *Policy Sciences* 36 (2): 175–95.

Halewood, Michael. 2008. "Introduction to Special Section on A. N. Whitehead." *Theory, Culture and Society* 25 (4): 1–14.

Hanna, Susan S. 1999. "Foreword." In *Fish, Markets, and Fishermen: The Economics of Overfishing*, edited by Suzanne Ludicello, Michael L. Weber, and Robert Wieland, ix–xii. Washington DC: Center for Marine Conservation and Island Press.

Hanna, Susan S., Carl Folke, and Karl-Goran Maler, eds. 1996. *Rights to Nature: Ecological, Economic, Cultural and Political Principles of Institutions for the Environment*. Washington DC: Island Press.

Haraway, Donna. 1988. "Situated Knowledges: The Science Question in Feminism and the Privilege of Partial Perspective." *Feminist Studies* 14 (3): 575–600.

Hardin, Garrett. 1968. "The Tragedy of the Commons." *Science* 162: 1243–48.

———. 1998. "The Feast of Malthus. Living within Limits." *Social Contract*, 181–87.

Hardt, Michael. 2010. "Two Faces of Apocalypse: A Letter from Copenhagen." *Polygraph* 22: 265–74.

Hardt, Michael, and Antonio Negri. 2009. *Commonwealth*. Cambridge MA: Harvard University Press.

Harvey, David. 1974. "Population, Resources and the Ideology of Science." *Economic Geography* 50 (3): 256–77.

———. 1996. *Justice, Nature and the Geography of Difference*. New York: Blackwell Publishers.

Heynen, Nik, James McCarthy, Scott Prudham, and Paul Robbins, eds. 2007. *Neoliberal Environments: False Promises and Unnatural Consequences*. London: Routledge.

Heynen, Nik, and Paul Robbins. 2005. "The Neoliberalization of Nature: Governance, Privatization, Enclosure and Valuation." *Capitalism Nature Socialism* 16 (1): 5–8.

Higgins, Vaughan, Jacqui Dibden, and Chris Cocklin. 2008. "Neoliberalism and Natural Resource Management: Agri-Environmental Standards and the Governing of Farming Practices." *Geoforum* 39 (5): 1776–85.

Hildyard, Nicholas, Larry Lohmann, Sarah Sexton, and Simon Fairlie. 1995. "Reclaiming the Commons." *The Corner House*. http://www.thecornerhouse.org.uk/resource/reclaiming-commons.

Holloway, John. 2005. "Zapatismo Urbano." *Humboldt Journal of Social Relations* 29: 168–78.

———. 2010. *Crack Capitalism*. New York: Pluto Press.

Hutton, William. 2013. "Why Do We Have to Trawl for the Facts about Britain and the EU?" *The Guardian*, June 2.

Hyde, Lewis. 2010. *Common as Air: Revolution, Art, and Ownership*. New York: MacMillan.

Illich, Ivan. 1983. "Silence is a Commons." *CoEvolution Quarterly*. 40: 5–9.

Ingold, Timothy. 1986. *The Appropriation of Nature. Essays on Human Ecology and Social Relations*. Manchester: Manchester University Press.

——. 2000. *Perception of the Environment Essays on Livelihood, Dwelling and Skill*. London: Routledge.

Jeffrey, Alex, Colin McFarlane, and Alex Vaseduvan. 2012. "Rethinking Enclosure: Space, Subjectivity and the Commons." *Antipode* 44 (4): 1247–67.

Jentoft, Svein. 2000. "The Community: A Missing Link of Fisheries Management." *Marine Policy* 24 (1): 53-59.

Jentoft, Svein, and Bonnie McCay. 1995. "User Participation in Fisheries Management. Lessons Drawn from International Experiences." *Marine Policy* 19 (3): 227–46.

——. 1996. "From the Bottom up: Participatory Issues in Fisheries Management." *Society and Natural Resources* 9: 237–50.

Joe. 2009. Interview with author, July 9. County Kerry.

Kaika, Maria. 2003. "Constructing Scarcity and Sensationalising Water Politics: 170 Days That Shook Athens." *Antipode* 35 (5): 919–54.

Kenis, Anneline, and Matthias Lievens. 2014. "Searching for 'the Political' in Environmental Politics" 23 (4): 1–18.

Khalilian, Setarah, Rainer Froese, Alexander Proelss, and Til Requate. 2010. "Designed for Failure: A Critique of the Common Fisheries Policy of the European Union." *Marine Policy* 34: 1178–82.

Klein, Naomi. 2014. *This Changes Everything: Capitalism vs. The Climate*. New York: Simon and Schuster.

Klooster, Dan. 2010. "Standardising Sustainable Development? The Forest Stewardship Council's Plantation Policy Review Process as Neoliberal Environmental Governance." *Geoforum* 41: 117–29.

Kolbert, Elizabeth. 2014. *The Sixth Extinction: An Unnatural History*. London: Bloomsbury.

Kompas, Tom, and Quentin R. Grafton. 2004. *Uncertainty and the Active Adaptive Management of Marine Reserves*. International and Development Economics Paper 04-2. Canberra: Australian National University.

Kooiman, Jan, Maarten Bavinck, Ratana Chuenpagdee, Robin Mahon, and Roger Pullin. 2008. "Interactive Governance and Governability: An Introduction." *The Journal of Transdisciplinary Environmental Studies* 7 (1): 1–11.

Kooiman, Jan, Svein Jentoft, Maarten Bavinck, and Roger Pullin, eds. 2005. *Fish for Life: Interactive Governance for Fisheries*. Amsterdam: Amsterdam University Press.

Larner, Wendy. 2000. "Neo-liberalism: Policy, Ideology, Governmentality." *Studies in Political Economy* 63: 5–25.

———. 2003. "Neoliberalism?" *Environment and Planning D* 21 (5): 509–12

Lars, Gulbrandsen. 2009. "The Emergence and Effectiveness of the Marine Stewardship Council." *Marine Policy* 33 (4): 654–60.

Latour, Bruno. 1993. *We Have Never Been Modern.* Harvester Wheatsheaf: Herefordshire.

———. 2005. *Reassembling the Social: An Introduction to Actor-Network-Theory.* Oxford: Oxford University Press.

———. 2005. "What is Given in Experience?" *Boundary* 32: 223–36.

Law, John. 2004. "And if the Global Were Small and Noncoherent? Method, Complexity, and the Baroque." *Environment and Planning D* 22 (1): 13–26

Lazzarato, Maurizio. 2009. "Neoliberialism in Action: Inequality, Insecurity and the Reconstitution of the Social." *Theory, Culture and Society* 26 (6): 109–33.

Leach, Melissa. 2008. "Pathways to Sustainability in the Forest? Misunderstood Dynamics and the Negotiation of Knowledge, Power, and Policy." *Environment and Planning A* 40 (8): 1783–95.

Leach, Melissa, Gerald Bloom, Adrian Ely, Paul Nightingale, Ian Scoones, Esha Shah, and Adrian Smith. 2007. *Understanding Governance: Pathways to Sustainability.* Brighton, UK: STEPS Centre, University of Sussex.

Lee, Sam. 2009. Interview with author, October 9. Galway.

Lemke, Thomas. 2001. "'The Birth of Bio-Politics'-Michel Foucault's Lecture at the College de France on Neo-Liberal Governmentality." *Economy and Society* 30 (2): 190–207.

———. 2002. "Foucault, Governmentality, and Critique." *Rethinking Marxism* 14: 49–64.

———. 2010. "Beyond Foucault. From Biopolitics to the Government of Life." In *Governmentality: Current Challenges and Future Issues*, edited by Ulrich Brockling and Susanne Krasman, 165–84. London: Routledge.

———. 2011a. *Bio-Politics: An Advanced Introduction.* New York and London: New York University Press.

———. 2011b. "Critique and Experience in Foucault." *Theory, Culture and Society* 28 (4): 26–47.

Leonardi, Emmanuele. 2012. "Biopolitics of Climate Change: Carbon Commodities, Environmental Profanations, and the Lost Innocence of Use-Value." Master's thesis, University of Western Ontario.

Lilley, Sasha, David McNally, and Eddie Yuen. 2012. *Catastrophism: The Apocalyptic Politics of Collapse and Rebirth.* Oakland CA: PM Press.

Linebaugh, Peter. 2008. *The Magna Carta Manifesto: Liberties and Commons for All.* Berkeley: University of California Press.

———. 2011. "Enclosure from the Bottom up." *Radical History Review* 108: 11–27.

———. 2012. *Ned Ludd & Queen Mab. Machine-Breaking, Romanticism, and the Several Commons of 1811–12.* Retort Pamphlet Series. Oakland CA: PM Press.

Linebaugh, Peter, and Marcus Rediker. 2000. *The Many-Headed Hydra: Sailors, Slaves, Commoners, and the Hidden History of the Revolutionary Atlantic*. Boston: Beacon Press.
Li, Tania M. 2002. "Engaging Simplifications: Community-Based Resource Management, Market Processes and State Agendas in Upland Southeast Asia." *World Development* 30 (2): 265–83.
———. 2005. "Beyond 'the State' and Failed Schemes." *American Anthropologist* 107: 383–94.
———. 2006. *Neo-Liberal Strategies of Government through Community: The Social Development Program of the World Bank in Indonesia*. Institute for International Law and Justice. https://tspace.library.utoronto.ca/bitstream/1807/67415/1/2006-2-GAL-Li-final-web.pdf.
Lockie, Stewart. 1999. "The State, Rural Environments, and Globalisation: 'Action at a Distance' via the Australian Landcare Program." *Environment and Planning A* 31 (4): 597–611.
Lohmann, Larry. 2009. "Neoliberalism and the Calculable World: The Rise of Carbon Trading." In *Upsetting the Offset: The Political Economy of Carbon Markets*, 25–40. London: Mayfly Books.
Ludwig, Donald, Ray Hilborn, and Carl Waters. 1993. "Uncertainty, Resource Exploitation, and Conservation: Lessons from History." *Science* 260: 17–36.
Mackenzie, Fiona D. 2010. "A Common Claim: Community Land Ownership in the Outer Hebrides, Scotland." *International Journal of the Commons* 4 (1): 319–44.
Mac Laughlin, John. 2010. *Troubled Waters: A Social and Cultural History of Ireland's Sea Fisheries*. Dublin: Four Courts Press.
Malthus, Thomas R. 1803. *An Essay on the Principle of Population*. 2nd edition. London: J. Johnson.
———. 2013. *An Essay on the Principle of Population*. Vol. 1. Cosimo Inc.
Mansfield, Becky. 2004. "Neoliberalism in the Oceans: 'Rationalisation', Property Rights, and the Commons Question." *Geoforum* 35: 313–26.
———. 2006. "Assessing Market-Based Environmental Policy Using a Case Study of North Pacific Fisheries." *Global Environmental Change* 16 (1): 29–39.
———. 2007a. "Property, Markets, and Dispossession: The Western Alaska Community Development Quota as Neoliberalism, Social Justice, Both, and Neither." *Antipode* 39 (3): 479–99.
———.2007b. "Articulation between Neoliberal and State-Oriented Environmental Regulation: Fisheries Privatization and Endangered Species Protection." *Environment and Planning A* 39: 1926–42.
March, Hug, and Thomas Purcell. 2014. "The Muddy Waters of Financialisation and New Accumulation Strategies in the Global Water Industry: The Case of AGBAR." *Geoforum* 53: 11–20.
Marcus, George E. 1995. "Ethnography in / of the World System: The Emergence of Multi-Sited Ethnography." *Annual Review of Anthropology* 24: 95–117.

Marine Stewardship Council. 2010. "Marine Stewardship Council." June 1. http://www.msc.org/newsroom/news/msc-responds-to-questions-about-antarctic-krill-certification.

———. 2011. *Get Certified! A Practical Guide to the Marine Stewardship Council's Fishery Certification Process.* Marine Stewardship Council. http://www.msc.org/documents/get-certified/fisheries/MSC_Get-certified_FINAL_lowres.pdf.

Marx, Karl. (1867) 1990. *Capital Volume I.* London: Penguin Books/New Left Review.

Marzac, Robert. 2011. "Energy Security: The Planetary Fulfillment of the Enclosure Movement." *Radical History Review* 109: 83–100.

Mathews, Freya. 1999. "Letting the World Grow Old: An Ethos of Countermodernity." *Worldviews: Global Religions, Culture, and Ecology* 32 (2): 119–37.

Mayhew, Jonathan R. 2014. *Malthus: The Life and Legacies of an Untimely Prophet.* Cambridge MA: Harvard University Press.

McCann, Charles R. 2004. *Individualism and the Social Order: The Social Element in Liberal Thought.* New York: Psychology Press.

McCarthy, Dave. 2009. Interview with author, May 28, Haulbowline, Cork.

McCarthy, James. 2005. "Commons as Counterhegemonic Projects." *Capitalism Nature Socialism* 16: 9–24.

McCarthy, James, and Scott Prudham. 2004. "Neoliberal Nature and the Nature of Neoliberalism." *Geoforum* 35 (3): 275–83.

McCay, Bonnie J. 1990. "The Culture of the Commoners. Historical Observations on Old and New World Fisheries." In *The Question of the Commons: The Culture and Ecology of Communal Resources*, edited by Bonnie J. McCay and James Acheson, 195–216. Tucson: University of Arizona Press.

———. 1996. "Common and Private Concerns." In *Rights to Nature: Ecological, Economic, Cultural, and Political Principles of Institutions for the Environment*, edited by Susan S. Hanna, Carl Folke, and Karl-Göran Mäler. Covelo, California: Island Press.

———. 2004. "ITQ and Community: An Essay on Environmental Governance." *Agricultural and Resource Economics Review* 33 (2): 162–70.

McMichael, Philip. 2011. "The Food Regime in the Land Grab: Articulating 'global Ecology' and Political Economy." Presented at the International Conference on Global Land Grabbing, University of Sussex, Brighton.

Mehta, Leyla, ed. 2013. *The Limits to Scarcity: Contesting the Politics of Allocation.* London: Earthscan.

Merchant, Carolyn. 1980. *The Death of Nature: Women, Ecology and the Scientific Revolution.* New York: Harper Collins.

Midnight Notes. *New Enclosures.* Jamaica Plain, MA: Midnight Notes, 1990.

Mies, Maria. 1998. *Patriarchy and Accumulation on a World Scale: Women in the International Division of Labour.* London: Zed Books.

Mies, Maria, and Veronika Bennholdt-Thomsen. 1999. *The Subsistence Perspective: Beyond the Globalised Economy.* Spinifex Press.

———. 2001. "Defending, Reclaiming and Reinventing the Commons." *Canadian Journal of Development Studies/Revue Canadienne D'études Du Développement* 22 (4): 997–1023.
Mikalsen, Knut, and Svein Jentoft. 2001. "From User-Groups to Stakeholders? The Public Interest in Fisheries Management." *Marine Policy* 25: 281–92.
Mol, Arthur P. J., and Gert Spaargen. 2000. "Ecological Modernisation Theory in Debate: A Review." *Environmental Politics* 9: 17–49.
Moore, Jason W. 2003. "Capitalism as World-Ecology. Braudel and Marx on Environmental History." *Organization & Environment* 16 (4): 431–58.
———. 2014a. "The Capitalocene, Part I: On the Nature and Origins of Our Ecological Crisis." Unpublished paper, Fernand Braudel Center, Binghamton University. http://www.jasonwmoore.com/uploads/The_Capitalocene__Part_I__June_2014.pdf.
———. 2014b. "The Capitalocene, Part II: Abstract Social Nature and the Limits to Capital." Unpublished paper. Binghamton NY: Fernand Braudel Center, Binghamton University.
Morton, Timothy. 2008a. "Ecologocentrism: Unworking Animals." *SubStance* 37: 73–96.
———. 2008b. "John Clare's Dark Ecology." *Studies in Romanticism* 47 (2): 179–93.
Mosse, David. 1997. "The Symbolic Making of a Common Property Resource: History, Ecology and Locality in Tank-Irrigated South India." *Development and Change* 28 (3): 467–504.
Murphy, Bill. Interview with author, August 27. Cork.
Murphy, Joseph. 2000. "Ecological Modernisation: Editorial." *Geoforum* 31: 1–8.
Murray, Grant, Teresa Johnson, Bonnie J. McCay, Kevin St. Martin, and Satsuki Takahashi. 2010. "Cumulative Effects, Creeping Enclosure, and the Marine Commons of New Jersey." *International Journal of the Commons* 4 (1): 367–89.
Nealon, Jeffrey Thomas. 2008. *Foucault Beyond Foucault: Power and Its Intensifications since 1984*. Palo Alto CA: Stanford University Press.
Neeson, Jeanette. 1996. *Commoners: Common Right, Enclosure and Social Change in England, 1700–1820*. Cambridge, UK: Cambridge University Press.
Nightingale, Andrea. 2013. "Fishing for Nature: The Politics of Subjectivity and Emotion in Scottish Inshore Fisheries Management." *Environment and Planning A* 45 (10): 2362–78.
Nolan, John. 2009. Interview with author, May 13. Castletownbere.
Olsson, Per, Lance H. Gunderson, Steve R. Carpenter, Paul Ryan, Louis Lebel, Carl Folke, and Crawford S. Holling. 2006. "Shooting the Rapids: Navigating Transitions to Adaptive Governance of Social-Ecological Systems." *Ecology and Society* 11 (1).
Ong, Aihwa, and Stephen J. Collier, eds. 2005. *Global Assemblages: Technology, Politics, and Ethics as Anthropological Problems*. Oxford: Blackwell Publishers.
Opitz, Sven. 2011. "Government Unlimited. The Security Dispositif of Illiberal Governmentality." In *Governmentality: Current Challenges and Future Issues*,

edited by Ulrich Bröckling, Susanne Krasmann, and Thomas Lemke, 93–114. New York: Routledge.
O'Reilly, Maeve. 2009. Interview with author, September 3. Cork.
Organisation for Economic Co-operation and Development. 2011. *Towards Green Growth*. OECD Green Growth Studies. Paris: OECD Publishing.
Österblom, Henrik, Michael Sissenwine, David Symes, Martina Kadin, Tim Daw, and Carl Folke. 2011. "Incentives, Social-Ecological Feedbacks and European Fisheries." *Marine Policy* 35: 568–74.
Ostrom, Elinor. 2000a. "Collective Action and the Evolution of Social Norms." *Journal of Economic Perspectives* 14 (3): 137–58.
———. 2000b. "Private and Common Property Rights." In *Encyclopedia of Law and Economics, Vol. II: Civil Law and Economics*, 332–79. Ghent, Belgium: University of Ghent.
———. 2008a. *Governing the Commons: The Evolution of Institutions for Collective Action*. Cambridge, UK: Cambridge University Press.
———. 2008b. *Institutions and the Environment*. 28. IEA Economic Affairs.
Ostrom, Elinor, Thomas Dietz, Nives Dolsak, Paul Stern, Susan Stonich, and Elke Weber, eds. 2002. *The Drama of the Commons*. Washington DC: National Academy.
Ostrom, Elinor, Roy Gardner, and James Walker. 1994. *Rules, Games, and Common-Pool Resources*. University of Michigan Press.
Ostrom, Elinor, and Edella Schlager. 1992. "Property-Rights Regimes and Natural Resources: A Conceptual Analysis." *Land Economics* 68 (3): 249–62.
O'Sullivan, Jean. 2009. Interview with author, June 30. Dublin.
Owens, Gerry. 2009. Interview with author, May 15. Castletownbere.
Palsson, Gisli. 1991. *Coastal Economies, Cultural Accounts. Human Ecology and Icelandic Discourse*. Manchester UK: Manchester University Press.
———. 1993. *Beyond Boundaries. Understanding, Translation and Anthropological Discourse*. Oxford: Berg.
Papadopoulos, Dimitris. 2010a. "Alter-Ontologies: Towards a Constituent Politics in Technoscience." *Social Studies of Science* 41: 177–201.
———. 2010b. "Insurgent Posthumanism." *Ephemera: Theory & Politics in Organization* 10: 134–51.
———. 2014. "Politics of Matter: Justice and Organization in Technoscience." *Social Epistemology: A Journal of Knowledge, Culture and Policy* 28 (1): 70–85.
Papadopoulos, Dimitris, and Niamh Stephenson. 2006. *Analysing Everyday Experience: Social Research and Political Change*. Hampshire and New York: Palgrave MacMillan.
Papadopoulos, Dimitris, Niamh Stephenson, and Vassilis Tsianos. 2008. *Escape Routes: Control and Subversion in the 21st Century*. London: Pluto Press.
Parenti, Christian. 2011. *Tropic of Chaos: Climate Change and the New Geography of Violence*. New York: Nation Books.

Paterson, B., M. Isaacs, M. Hara, A. Jarre, and C.L. Moloney. 2010. "Transdisciplinary Co-Operation for an Ecosystem Approach to Fisheries: A Case Study from the South African Sardine Fishery." *Marine Policy* 34: 782–94.
Peace, Adrian. 2001. *A World of Fine Difference: The Social Architecture of a Modern Irish Village*. Dublin: University College Dublin Press.
Peck, Jamie. 2004. "Geography and Public Policy: Constructions of Neoliberalism." *Progress in Human Geography* 28 (3): 392–405.
Peck, Jamie, Nik Theodore, and Neil Brenner. 2010. "Postneoliberalism and Its Malcontents." *Antipode* 41 (S1): 94–116.
Peck, Jamie, and Adam Tickell. 2002. "Neoliberalizing Space." *Antipode* 34 (3): 380–404.
Peet, Richard, and Michael Watts. 1996. *Liberation Ecologies. Environment, Development, Social Movements*. London: Routledge.
Pellizzoni, Luigi. 1999. "Reflexive Modernisation and Beyond: Knowledge and Value in the Politics of Environment and Technology." *Theory, Culture and Society* 16 (4): 99–125.
———. 2011. "Governing through Disorder: Neoliberal Environmental Governance and Social Theory." *Global Environmental Change* 21: 795–803.
Perelman, Michael. 2001. "The Secret History of Primitive Accumulation and Classical Political Economy." *The Commoner* 2.
Perry, R. Ian, and Rosemary E. Ommer. 2010. "Introduction: Coping with Global Change in Marine Social-Ecological Systems." *Marine Policy* 34 (4): 739–41.
Pirages, Dennis, and Ken Cousins, eds. 2005. *From Resource Scarcity to Ecological Security*. Cambridge MA: MIT Press.
Ponte, Stefano, and Emmanuelle Cheyns. 2013. "Voluntary Standards, Expert Knowledge and the Governance of Sustainability Networks." *Global Networks* 13 (4): 459–77.
Poseidon Aquatic Resource Management Ltd. & Pew Environment Group. 2010. *Financial Instrument for Fisheries Guidance 2000–2006*. Hampshire UK.
Puig de la Bellacasa, Maria. 2010. "Ethical Doings in Naturecultures." *Ethics, Place and Environment* 13: 151–69.
———. 2011. "Matters of Care in Technoscience: Assembling Neglected Things." *Social Studies of Science* 41 (1): 85–106.
———. 2012. "'Nothing Comes without Its World': Thinking with Care." *The Sociological Review* 60 (2): 197–216.
Rabinow, Paul. 2003. *"Anthropos Today." Reflections on Modern Equipment*. Princeton NJ: Princeton University Press.
Rabinow, Paul, and Nikolas Rose. 2006. "Biopower Today." *Biosocieties* 1: 195–217.
Ranciere, Jacques. 1998. *Disagreement: Politics and Philosophy*. University of Minnesota Press.
Read, Jason. 2009. "A Genealogy of Homo-Economicus: Neoliberalism and the Production of Subjectivity." *Foucault Studies* 6: 25–36.

Robertson, Morgan M. 2006. "The nature that capital can see: science, state, and market in the commodification of ecosystem services." *Environment and Planning D* 24 (3): 367.

Rogers, Raymond. 1995. *The Oceans Are Emptying: Fish Wars and Sustainability*. Montreal: Black Rose Books.

Rose, Deborah Bird. 1996. *Nourishing Terrains: Australian Aboriginal Views of Landscape and Wilderness*. Australian Heritage Commission.

———. 2004. *Reports from a Wild Country: Ethics for Decolonisation*. Sydney: UNSW Press.

———. 2013. "Val Plumwood's Philosophical Animism: Attentive Inter-Actions in the Sentient World." *Environmental Humanities* 3: 93–109.

Rose, Nikolas. 2006. "Governing 'Advanced' Liberal Democracies." In *The Anthropology of the State: A Reader*, edited by Akhil Gupta and Aradhana Sharma, 144–62. Oxford: Blackwell Publishing.

Rose, Nikolas, and Peter Miller. 1995. "Political Power beyond the State: Problematics of Government. *The British Journal of Sociology* 43 (2): 173–205.

Rose, Philip. 2002. *On Whitehead*. Belmont CA: Wadsworth.

Ross, Eric B. 1998. *The Malthus Factor: Poverty, Politics and Population in Capitalist Development*. London: Zed Books.

Rowe, Jonathan. 2001. "The Hidden Commons." *Yes! Magazine* 18 (summer). http://www.yesmagazine.org/issues/reclaiming-the-commons/the-hidden-commons.

St. Martin, Kevin. 2000. "Mapping Economic Diversity in the First World: The Case of Fisheries." *Environment and Planning A* 37: 959–79.

———. 2005. "Disrupting Enclosure in New England Fisheries." *Capitalism Nature Socialism* 16: 63–80.

———. 2007. "The Difference That Class Makes: Neoliberalisation and Non-Capitalism in the Fishing Industry of New England." *Antipode* 39: 528–49.

———. 2009. "Toward a Cartography of the Commons: Constituting the Political and Economic Possibilities of Place." *Professional Geographer* 61: 493–507.

Schmitter, Philippe. 2002. "Participation in Governance Arrangements: Is There Any Reason to Expect It Will Achieve 'sustainable and Innovative Policies in a Multi-Level Context'?." In *Participatory Governance: Political and Societal Implications*, edited by Jürgen Grote and Bernard Gbikipi. Opladen: Leske and Budrich.

Schneider, Mindi, and Philip McMichael. 2010. "Deepening, and Repairing, the Metabolic Rift." *The Journal of Peasant Studies* 37 (3): 461–84.

Scott, James C. 1998. *Seeing Like a State: How Uncertain Schemes to Improve the Human Condition Have Failed*. New Haven: Yale University Press.

Shaviro, Steven. 2011. "The 'Bitter Necessity' of Debt: Neoliberal Finance and the Society of Control." *Concentric: Literary and Cultural Studies* 37 (1): 73–82.

Sheehan, Ebbie. 2009. "Open Letter." *Irish Skipper*, October.

Shiva, Vandana. 2010. "Resources." In *The Development Dictionary*, edited by Wolfgang Sachs, 228–43. Zed Books.

Silver, Jennifer J., and Roberta Hawkins. 2014. "'I'm Not Trying to Save Fish, I'm Trying to Save Dinner': Media, Celebrity and Sustainable Seafood as a Solution to Environmental Limits." *Geoforum*.

Sissenwine, Michael, and David Symes. 2007. "Reflections on the Common Fisheries Policy: Report to the General Directorate for Fisheries and Maritime Affairs of the European Commission." European Commission.

Smith, Brid. 2009. Interview with author, June 1. Clonakilty, County Cork.

Smith, Laila. 2004. "The Murky Waters of the Second Wave of Neoliberalism: Corporatization as a Service Delivery Model in Cape Town." *Geoforum* 35 (3): 375–93.

Springer, Simon. 2012. "Neoliberalism as Discourse: Between Foucauldian Political Economy and Marxian Poststructuralism." *Critical Discourse Studies* 9 (2): 133–47.

Starhawk. 1982. "Appendix A." In *The Burning Times: Notes on a Crucial Period of History, Dreaming the Dark*. Boston: Beacon Press.

Stengers, Isabelle. 2008. "Experimenting with Refrains: Subjectivity and the Challenge of Escaping Modern Dualism." *Subjectivity* 22: 38–59.

Stephenson, Niamh. 2003. "Interrupting Neo-Liberal Subjectivities." *Continuum: Journal of Media & Cultural Studies* 17: 135–46.

Sullivan, Sian. 2013. "Banking Nature? The Spectacular Financialisation of Environmental Conservation." *Antipode* 45 (1): 198–217.

Swyngedouw, Erik. 1999. "Governance Innovation and the Citizen: The Janus Face of Governance-beyond-the-State." *Urban Studies* 42: 1991–2006.

———. 2004. "Globalisation or 'glocalisation'? Networks, Territories and Rescaling." *Cambridge Review of International Affairs* 17 (1): 25–48.

———. 2007. "Impossible Sustainability and the Post-Political Condition." In *The Sustainable Development Paradox: Urban Political Economy in the United States and Europe*, edited by Rob Krueger and David Gibbs, 13–40. New York: Guilford Press.

———. 2010. "Apocalypse Forever? Post-Political Populism and the Spectre of Climate Change." *Theory, Culture, and Society* 27 (2–3): 213–32.

Symes, David. 2005. "Altering Course: Future Directions for Europe's Fisheries Policy." *Fisheries Research* 71: 259–65.

———. 2007. "Fisheries Management and Institutional Reform: A European Perspective." *International Council for the Exploration of the Sea*.

———. 2009. "Reform of the European Union's Common Fisheries Policy: Making Fisheries Management Work." *Fisheries Research* 100 (2): 99–102.Terranova, Tiziana. 2009. "Another Life. The Nature of Political Economy in Foucault's Genealogy of Biopolitics." *Theory, Culture and Society* 26 (6): 234–62.

Thompson, E. P. 1993. *Customs in Common. Studies in Traditional Popular Culture.* New York: The New Press.

Tronto, Joan. 1993. *Moral Boundaries: A Political Argument for an Ethic of Care.* New York: Routledge.

Tully, Oliver. 2004. "Integration of Biology and Management in Lobster Fisheries." In *The Biology and Management of Clawed Lobsters (Homarus Gammarus L.) in Europe*, edited by Oliver Tully. Fisheries Resource Series 1–4. Bord Iascaigh Mhara.

Tully, Oliver, A. O'Leary, A. McCarthy, V. O'Donovan, and D. Nee. 2006. *The Lobster (Homarus Gammarus L.) Fishery: Analysis of the Resource in 2004–2005*. Fisheries Resource. Bord Iaschaigh Mhara.

Turnhout, Esther, Katja Neves, and Elisa de Lijster. 2014. "'Measurability' in Biodiversity Governance: Knowledge, Transparency and the Intergovernmental Science Policy Platform on Biodiversity and Environments (IPBES)." *Environment and Planning A* 46: 581–97.

Uexkull, Jakob Von. 1957. "A Stroll through the World of Animals and Men." In *Instinctive Behavior: The Development of a Modern Concept*, edited by Claire H. Schiller. New York: International Universities Press.

Van der Ploeg, Jan Douwe. 2008. *The New Peasantries Struggles for Autonomy and Sustainability in an Era of Empire and Globalisation*. London: Earthscan.

Viveiros de Castro, Eduardo. 1998. "Cosmological Deixis and Amerindian Perspectivism." *The Journal of the Royal Anthropological Institute* 4 (3): 469–88.

Walton, Izaak. 2000. *The Compleat Angler*. Oxford: Oxford University Press.

Warner, Rosalind. 2010. "Ecological Modernisation Theory: Towards a Critical Ecopolitics of Change?" *Environmental Politics* 19: 538–56.

Whelan, Bill. 2009. Interview with author, June 5. Dublin.

Whitehead, Alfred North. 1967. *Science and the Modern World*. New York: The Free Press.

Wiber, Melanie, Fikret Berkes, Anthony Charles, and John Kearney. 2004. "Participatory Research Supporting Community-based Fishery Management." *Marine Policy* 28 (6): 459–68.

Whooley, Jason. 2011. *The Irish Skipper*.

World Bank. 1992. *World Development Report 1992: Development and Environment*. Washington DC: World Bank, 1992.

Winch, Donald. 2013. *Malthus: A Very Short Introduction*. Oxford: Oxford University Press.

Young, Arthur. 1897. *A Tour In Ireland, 1776–79*. London: Cassell & Company.

Yuen, Eddie. 2012. "The Politics of Failure Have Failed: The Environmental Movement and Catastrophism." In *Catastrophism: The Apocalyptic Politics of Collapse and Rebirth*, edited by Sasha Lilley, David McNally, and Eddie Yuen, 15–43. Oakland CA: PM Press.

Zibechi, Raul. 2012. *Territories in Resistance: A Cartography of Latin American Social Movements*. Oakland CA: AK Press.

Index

Milton Keynes UK
Ingram Content Group UK Ltd.
UKHW030856040824
446415UK00012B/156